■

TECHNOLOGY AND THE LIMITATION OF INTERNATIONAL CONFLICT

■

THE JOHNS HOPKINS
FOREIGN POLICY INSTITUTE

The Johns Hopkins Foreign Policy Institute (FPI) was founded in 1980 and serves as the research center for the School of Advanced International Studies (SAIS) in Washington, D.C. The FPI is a meeting place for SAIS faculty members and students as well as for government analysts, policymakers, diplomats, journalists, business leaders, and other specialists in international affairs. In addition to conducting research on policy-related international issues, the FPI sponsors conferences, seminars, and roundtables.

Current research activities at the FPI span the complete spectrum of American foreign policy and international affairs. In the project on "U.S.-Soviet Relations: An Agenda for the Future" FPI fellows and outside experts evaluate the recent evolution of U.S.-Soviet relations and develop original ideas for increased super-power cooperation in new and established areas. Through the project on foreign policy consensus the FPI seeks to advance the national dialogue on central issues of U.S. foreign policy: specific recommendations are prepared by selected experts and endorsed by a bipartisan commission for wide distribution in the policy community.

Other current FPI programs examine the impact of the landmark Goldwater-Nichols defense reorganization act; the relation of arms control to force structure and military-political doctrine; the politics of international terrorism; the role of the media in foreign policy; American and Soviet national security policymaking; and other leading international issues. These programs are usually directed by FPI fellows.

FPI publications include the *SAIS Review*, a semiannual journal of foreign affairs, which is edited by SAIS students; the FPI Papers in International Affairs, a monograph series designed to make public the best and most cogent scholarly work on foreign policy and defense issues; the FPI Policy Briefs, a series of analyses of immediate or emerging foreign-policy issues; the FPI Case Studies, a series designed to teach analytical negotiating skills; and the FPI Policy Consensus Reports, which present recommendations on a series of critical foreign policy issues.

For additional information regarding FPI activities, write to: FPI Publications Program, School of Advanced International Studies, The Johns Hopkins University, 1619 Massachusetts Avenue, N.W., Washington, D.C. 20036-2297.

FPI PAPERS IN INTERNATIONAL AFFAIRS

■

TECHNOLOGY AND THE LIMITATION OF INTERNATIONAL CONFLICT

Barry M. Blechman, Editor

with the assistance of

David K. Boren

■

FPI

Foreign Policy Institute
School of Advanced International Studies
The Johns Hopkins University
Washington, D.C.

Library of Congress Cataloging-in-Publication Data

Technology and the limitation of international conflict / Barry M. Blechman,
editor, with the assistance of David K. Boren.
    p.   cm.
(FBI papers in international affairs)
1. Arms control--Verification. 2. Military surveillance. 3. Nuclear weapons-
-Safety measures. 4. Nuclear warfare--Safety measures. I. Blechman, Barry
M. II. Boren, David, 1941-   . III. Johns Hopkins University. Foreign Policy
Institute. IV. Series.
UA12.5.T44   1988   88-24459 CIP
327.1'74'028--dc19
ISBN 0-941700-43-7 (alk. paper).
ISBN 0-941700-42-9 (pbk. : alk. paper)

Distributed by arrangement with

UPA, Inc.
4720 Boston Way
Lanham, Md. 20706/(301) 459-3366

3 Henrietta Street
London, WC2E 8LU
England

ABOUT THE AUTHORS

Barry Blechman is a fellow of The Johns Hopkins Foreign Policy Institute and president of Defense Forecasts Inc., a research and analysis enterprise in Washington, D.C. He served as assistant director of the U.S. Arms Control and Disarmament Agency from 1977 to 1980 and is coauthor of *The Silent Partner: West Germany and Arms Control* (1988).

William S. Cohen is the senior Republican senator from Maine. He is vice chairman of the Select Committee on Intelligence and ranking Republican on the subcommittee on Projection Forces and Regional Defense of the Armed Services Committee.

Les Aspin is a Democratic representative from Wisconsin. He holds a master's degree from Oxford University and a Ph.D. in economics from the Massachusetts Institute of Technology. Representative Aspin has been chairman of the House Armed Services Committee since January 1985.

Ivan Oelrich is a research fellow at the Center for Science and International Affairs at Harvard University. His previous work at the Institute for Defense Analyses, has concentrated primarily on military research and development and arms control. He is the author of "Production Monitoring for Arms Control" in *Arms Control* (Spring 1988).

Victor Utgoff is a deputy director of the strategy, forces and resources division for the Institute for Defense Analyses. His current areas of research include remote sensors for confidence-building and policy for theater nuclear forces. He is coauthor of *The Fiscal and Economic Implications of SDI* (1986) among other publications.

Steven Fetter is an assistant professor in the School of Public Affairs at the University of Maryland, College Park. The chapter on tagging treat-limited items was completed while he was a research fellow at the Center for Science and International Affairs in the Kennedy School of Government, Harvard University. Dr. Fetter received an S.B. in physics from the Massachusetts Institute of Technology and a Ph.D. in energy and resources from the University of California, Berkeley. He has published articles on fission and fusion reactor safety, radioactive waste disposal, space policy and arms control verification.

Thomas Garwin is currently on the faculty of the University of Maryland, serving as the American coordinator for the Nuclear History Program, a multi-national program that sponsors research and discussions concerning the role of nuclear weapons in the evolution of international relations since World War II. He has worked at the office of the secretary of defense, The Brookings Institution, and the U.S. Office of Technology Assessment. He has also consulted for a variety of government agencies and non-profit organizations. Mr. Garwin holds an A.B. in history and a masters in public policy from Harvard University.

Franklin E. Walker is a senior scientist at Lawrence Livermore National Laboratory; he is currently assigned to the treaty verification program. Dr. Walker served as senior scientific advisor in the Department of State Bureau of Intelligence and Research for two an one-half years, during which time he visited the Conference on Disarmament and the U.S. mission in Geneva. He completed a survey of the chemical weapons monitoring instrumentation programs of many West European countries in September 1986.

Michael Krepon is a senior associate at the Carnegie Endowment for International Peace where he directs the commercial observation satellite and verification projects. He served in the Carter administration, directing defense projects and policy reviews at the Arms Control and Disarmament Agency. He previously worked on Capitol Hill as an assistant to Rep. Norman D. Dicks (D-Wash.). He is the author of *Strategic Stalemate: Nuclear Weapons and Arms Control in American Politics* (1984) and co-editor of *Verification and Compliance: A Problem-Solving Approach* (1988).

Paul B. Stares is currently a research associate in the Foreign Policy Studies Program at The Brookings Institution, Washington, D.C. He received his Ph.D. in politics from the University of Lancaster, England, where he was also research fellow at the Centre for the Study of Arms Control and International Security. Subsequently he became a Rockefeller Foundation international relations fellow and guest scholar at Brookings before assuming his present position. He is the author of numerous studies on military space issues including two books, *The Militarization of Space: U.S. Policy, 1945–1984* (1986) and *Space and National Security* (1987).

William H. Lewis is the director of security policy studies and professor of international relations at George Washington University. Dr. Lewis was a foreign service officer, having served as office director in the Bureaus of Political-Military and African Affairs. He was a member of the Presidential Task Force on Foreign Aid (1969–70), a senior fellow at The Brookings Institution (1974–75), and the Carnegie Endowment (1979), and a visiting professor at the University of Michigan (1965–66). He is the co-author of *Debacle: The American Failure in Iran* (1981).

Gerald W. Johnson is an adjunct professor of political science at the University of California, San Diego. He served as the assistant for atomic energy to the secretary of defense, 1961–63, in which capacity he was responsible for the safety of nuclear weapons and the introduction of permissive action links into the U.S. and NATO stockpiles, and as the personal representative of the secretary of defense to SALT II and the comprehensive test ban negotiations, 1977–79. He has written widely on nuclear energy and policy. He is coauthor of "The History of Strategic Arms Limitations" in *Bulletin of Atomic Scientists* (January 1984) and author of "Underground Nuclear Explosions and Seismology—A Cooperative Effort" in *The Vela Program, A Twenty-five Year Review of Basic Research*, Ann U. Kerr, ed. (1985)

Dan Caldwell is currently a professor of political science at Pepperdine University, Malibu, California, and senior research associate at the Center for International and Strategic Affairs at the University of California, Los Angeles. He received A.B., M.A., and Ph.D. degrees from Stanford University and a master's degree from the Fletcher School of Law and Diplomacy. Dr. Caldwell is the author of *American-Soviet Relations: From 1947 to the Nixon-Kissinger Grand Design and Grand Strategy* and *Soviet International Behavior and U.S. Policy Options*. He is now writing a book on the domestic politics of the SALT II treaty ratification debate.

Peter D. Zimmerman, a physicist, is a senior associate at the Carnegie Endowment for International Peace and director of its program on the Strategic Defense Initiative and arms control. He was a William C. Foster fellow at the Arms Control and Disarmament Agency from 1984–86. In that capacity he served as an adviser to the U.S. delegation to the Strategic Arms Reduction Talks in Geneva. He was professor of physics at Louisiana State University from 1974–86.

TABLE OF CONTENTS

FOREWORD

Technological change is generally agreed to affect the stability of the military balance and the risk of war. But the magnitude and direction of the effect has long been debated by military analysts and practitioners. The effects of scientific advances on these problems are often assumed to be invariably negative. The assumption stems primarily from the single most important advance in military technology of our time—the development of nuclear weapons. Ballistic missiles have compounded that perception by making strategic war as swift as it would be destructive. But the destructiveness of a war using such weapons does not necessarily make it more likely, or peace less stable—the contrary argument has at least as much weight. In principle, however, technology is neither necessarily "good" nor necessarily "bad" for stability and the risk of war. The challenge is to channel technological change in the right directions: technologies can be utilized to increase stability and reduce the likelihood of war, just as they can be used to strengthen military power or threaten surprise attack.

In fact, during the postwar period, technological advances have made important contributions to our ability to limit armaments and reduce the risk of war. Examples include controls on nuclear weapons, such as electronic locks and command systems, which are far more secure now than in the past. Photoreconnaissance satellites and other national technical means of intelligence represent a second example; these systems not only enable a nation's leader to have greater confidence in his or her knowledge of the capabilities and actions of potentially hostile countries but also make possible the far-reaching types of arms control agreements now being discussed by the United States and the Soviet Union.

New technologies potentially could be used in other ways to reduce the risk of international conflict and to facilitate arms limitations. To help uncover such new ideas and promote their development, The Johns Hopkins Foreign Policy Institute in 1985 organized the project on Technology and the Limitation of International Conflict. Cochaired by Senator William Cohen and Congressman Les Aspin, the project's executive committee included eighteen former senior defense and intelligence officials and experts in relevant scientific and engineering disciplines. The group examined a variety of proposals ranging from additional means of safeguarding nuclear weapons, to methods for avoiding international incidents in space, to a scheme for cooperative use of remote sensors to avoid surprise conventional attacks in Europe.

The committee gave special attention to technical means to help verify compliance with possible international arms control agreements. Technology cannot replace the political will that is necessary for the negotiation and implementation of stabilizing arms control agreements, but it can assist in alleviating concerns about the monitoring and verification of treaty compliance. Any sound international agreement must be adequately verifiable, but for reasons of security and propriety nations are sometimes reluctant to permit intrusive on-site inspections by foreign personnel. The need to satisfy both criteria suggests the possibility that technical devices might be used to reduce requirements for human inspections, thus facilitating the negotiation of arms control agreements that are under consideration now or may become possible in the future.

One major factor inhibiting the application of technology to promote international stability is the gap that frequently exists between the policymaking community and the world of engineering and technical expertise. The collaboration between policymakers and technical experts on this project was of significant importance. Once the executive committee had identified the most promising areas in which technology could assist in the reduction of international conflict, it commissioned experts with experience in the specific fields under consideration to examine and describe how the relevant technologies could be applied. The committee evaluated these analyses in a policy context to assess the potential practical utility of the concepts, measure the policy risks of excessive expectations or malfunctions, and, when relevant, determine the negotiability of the idea. Particular attention was given to how good ideas could support actual agreements.

The present volume is a compilation of some of the most promising means of using technology to reduce international conflict that were discussed by the committee. They merit the attention due any serious proposal whose purpose is to channel technology, one of our greatest assets, in a positive direction.

The Johns Hopkins Foreign Policy Institute is the research arm of the School of Advanced International Studies. The Project on Technology and the Limitation of International Conflict was made possible by grants from the Rockefeller Brothers Fund, the W. Alton Jones Foundation, Inc., and the Florence and John Schumann Foundation. We are grateful for their support.

Harold Brown

William Perry, former undersecretary of defense for research and engineering, now chairman of H & Q Technology Partners Inc.

Lt. General Brent Scowcroft, USAF ret., national security adviser, 1973–77, now vice chairman of Kissinger Associates

Lt. General Eugene Tighe, USAF ret., former director of the Defense Intelligence Agency

Victor Utgoff, former senior staff member of the National Security Council, now deputy director of the division on strategy, forces and resources at the Institute for Defense Analyses

Herbert York, former director of Defense Research and Engineering, now director of the Institute on Global Conflict and Cooperation

1

TECHNOLOGY AND THE LIMITATION OF INTERNATIONAL CONFLICT

BY WILLIAM S. COHEN AND LES ASPIN

Technology is a decisive force in contemporary American life. It elicits both hope for the future and dread that the future may not arrive. Either view, however, presupposes that technology is inherently "good" or "bad." In reality, technology is the neutral instrument of humanity. It can only be used for good or bad purposes; its impact will depend on our decisions.

Since the development of the atomic bomb in 1945, the link between technology and the continuing competition in nuclear and conventional armaments has contributed primarily to the idea that technology is inherently bad. Many of the most highly publicized technological advances have enhanced the destructive potential of modern weapon systems. The development of intercontinental ballistic missiles, the invention of multiple independently targeted reentry vehicles, and the great accuracy of modern missiles, among many other technological advances, fit this description.

Less well publicized but equally important, however, has been the application of new technologies to increase international stability and to improve the safety and security of nuclear weapons. Indeed, it is clear that in 1988 the arsenals of the nuclear powers are far less prone to accidents or to inadvertent or unauthorized uses than was the case thirty, or even twenty, years ago. Technological developments of all sorts have helped to lessen the danger of maintaining large nuclear arsenals. It is now possible to make modern missiles and the platforms upon which they are deployed less vulnerable to attack than their predecessors and, hence, to lessen pressure to launch them preemptively in a crisis. Modern command and control systems and electronic locking devices have greatly diminished the risk of unauthorized or premature launchings of nuclear weapons and reduced the possibility that nuclear war might begin in response to false warnings of an attack. Such safety measures as the use of inert conventional explosives for the triggering mechanisms of nuclear weapons have virtually eliminated legitimate fears of accidental detonations.

Photoreconnaissance and other types of warning and intelligence satellites enable governments to appraise confidently the size and characteristics

1

of potentially hostile military forces and to monitor closely their disposition and readiness for war. Effective surveillance is a source of stability
in crises and has reduced pressures to build incremental military capabilities based on unwarranted suspicions or mindless fears of other
nations' military might. Satellite-based reconnaissance systems also have
made possible negotiated restraints on armaments that would have been
unthinkable several decades ago.

Many novel ways of applying technology to promote international stability and to reduce the risk of war remain unexplored. Rapid technological
change sometimes makes it difficult to know how best to exploit the potential of new technologies in concrete proposals. In some cases, ideas for technological applications have kicked around for years in scientific or
engineering circles without coming to the attention of policymakers in positions to promote their development and use. In other cases, suspicions and
political tensions between nations prevent the implementation of technological proposals that require cooperative activities. In still other cases, there
is national resistance to new and different ways of doing things. When such
bureaucratic resistance does not exist, or is overcome, insufficient resources
can sometimes preclude the development of technological applications to
the point at which their utility can actually be demonstrated.

The essays in this volume represent the combined efforts of policymakers, technical experts, and scholars to identify promising technologies
that will promote international stability and close the gap between their
conception and application. Each of the technologies has been examined
to determine its potential for reducing the likelihood of war or the risks
inherent in the existence of large stocks of weapons of mass destruction
throughout the world. This report results from several years of research
into "Technology and the Limitation of International Conflict"—a project
sponsored by the Rockefeller Brothers Fund, the W. Alton Jones Foundation, and the Florence and John Schumann Foundation and conducted
under the auspices of The Johns Hopkins Foreign Policy Institute of the
School of Advanced International Studies.

The study group was organized in 1985 to bridge the gap between
members of the scientific/technical communities and policymakers in the
executive branch and the Congress. The executive committee, which we
cochaired, included former high-ranking U.S. defense and foreign policy
officials, retired military officers, and other senior individuals with scientific and technical backgrounds; the members of the committee are listed
at the end of this article. The committee guided the project's overall development by identifying broad areas of interest, finding qualified specialists
to develop specific proposals, and reviewing these individuals' research
methods and conclusions. Among the technical experts who prepared
analyses for the project were several with backgrounds in research at national laboratories, large corporations, and universities.

2

The individual authors alone bear responsibility for the contents of their articles. The committee certifies only that these are interesting ideas meriting further consideration.

Much of the project was devoted to studying methods by which compliance with arms control treaties could be verified more easily and confidently. Improving verification technologies could make possible the pursuit of increasingly ambitious agreements by reducing suspicions that one side or the other might cheat on an agreement and by reducing requirements for on-site human inspections—a technique for treaty verification that can be difficult to implement effectively.

The United States also has begun to rethink its position on on-site human inspections. Although, in principle, the right to inspect suspicious facilities any time, anywhere, is desirable, serious thought on the matter soon reveals certain problems with the approach. There would be, for example, concerns about the violation of private property rights and the compromise of commercial secrets. There exist, in addition, certain military and intelligence facilities in the United States to which Soviet personnel should not have access under any circumstances. For these reasons, and others, the application of technologies that might reduce requirements for human on-site inspections has gained added interest in recent years. In considering how to verify a chemical treaty, for example, a particular problem arises from the fact that since the end of World War II the variety of chemicals with lethal or otherwise injurious properties has grown significantly in number and extent of use. Many are regularly used for agricultural and industrial purposes. Inspection techniques have failed to keep pace with developments in chemical weapons technology and the methods that could be used to disguise chemical agents. Yet, progress in verification technology and inspection techniques has been made.

For the NATO-Warsaw Pact negotiations on conventional stability and force reductions in Europe to be successful, improved monitoring technologies are essential. The committee discussed a promising idea in this area. Victor Utgoff and Ivan Oelrich of the Institute for Defense Analyses have developed and analyzed a proposal for "Confidence Building with Unmanned Sensors in Central Europe"; their proposal is presented in chapter two.

The proposed monitoring system would include seismic acoustic sensors deployed in and around the garrisons on both sides of the East/West border. These sensors could detect preparations for attacks by armored divisions located near the potential battle zone. A second set of sensors, including riverine sonars and cameras on major bridges and roadways, would form a continuous detection "barrier" distant enough from the East/West border to allow early warning of the movement of divisions stationed in garrisons farther from the border. The proposed monitoring system would be tamperproof. Outright disablement of the sensors would itself be an important signal to prepare for a possible forthcoming attack. The use of encryption devices to code the collected information prior to transmission would ensure the accuracy of the data and forestall enemy deceptions.

The use of remote sensors to monitor the movement of heavy military equipment of both NATO and the Warsaw Pact on a continuing basis could supplement national means of intelligence and substantially increase each side's confidence in its ability to detect preparations for a surprise attack. During peacetime, the sensors would enable each side to develop an accurate picture of the normal military routines of the other. As a result, both sides would be better able to identify the changes in daily routines that would necessarily occur prior to an attack. During periods of increased tension, the sensors would complement existing monitoring systems, such as reconnaissance aircraft with side-looking radars that can view only specific areas. The proposed remote sensors, which have the advantage of providing constant and comprehensive observation, could assist in identifying the specific areas at which intelligence assets should be directed to look more closely.

Given the historic importance of surprise to successful attack, such measures have the potential to reduce significantly the likelihood of conventional attack. Moreover, the cooperative emplacement of these sensors by NATO and the Warsaw Pact would in itself constitute a major confidence-building measure. Utgoff and Oelrich have briefed U.S. government officials on this proposal, as well as European defense experts. The idea has been favorably received within the defense community and if endorsed by the NATO allies could be an important complement to a U.S.-Soviet arms control agreement reducing ground forces stationed in central Europe.

Another important verification concern is mobile missiles. Although the INF treaty concluded in 1987 side-stepped this problem by requiring the total elimination of mobile Soviet SS-20 and U.S. ground-launched cruise missiles, the ability of the superpowers to verify the numbers of mobile missiles deployed on each side could be especially important in the event of a future agreement on central strategic systems that reduced, but did not abolish, such weapons. The committee has examined an innovative new concept for dealing with this problem—missile tags.

Two types of tags could be designed to enable either party to an agreement limiting mobile missiles to monitor the number of such missiles deployed by the other side. One method would entail marking the missiles of both states with a clear plastic tag containing small particles of light-reflecting material. A specially designed camera could then be used to photograph the tag from different directions and thereby provide a record of the precise number and location of reflections in the tag. The small size and the number of reflections would make these tags virtually impossible to reproduce. Challenge inspections would enable either side to ascertain whether untagged missiles, or missiles with false tags, were being deployed in violation of the agreement.

A second and more sophisticated tag would be an electronic chip attached to each missile. Each of these tags would emit signals verifying that it was the original tag and not a substitute. Electronic chips would have the additional advantage of being able to support a treaty limiting the number of missiles allowed in a certain geographical area by sending signals to satellites to verify that the missile had not entered a forbidden area. Safeguards could be included to prevent these tags from assisting in the targeting of the missiles by the other side's missiles. Potential applications of tags for monitoring restrictions on mobile missiles are discussed by Steven Fetter of the John F. Kennedy School of Government and Thomas Garwin of the MacArthur Foundation in chapter three, "Using Tags to Monitor Numerical Limits in Arms Control Agreements."

One area in which improved verification technologies could contribute directly to the conclusion of an agreement in the near future is in the field of chemical weapons. The threat of chemical warfare has risen dramatically in recent years. An increasing number of nations have obtained the means of building lethal chemical weapons, with perhaps fifteen to twenty now having operational stocks. Relatively recent uses of chemical agents by Iraq and Vietnam, and the Soviet Union's continued emphasis on chemical warfare in its military preparations, underscore the potential benefits of an agreement to ban these weapons.

As a step toward eliminating chemical weapons, Vice President George Bush tabled a "Draft Convention on the Prohibition of Chemical Weapons," on April 18, 1984, at the Geneva Conference on Disarmament (CD). The proposal calls for far-reaching restrictions on the production, storage, transportation, and destruction of lethal chemicals with weapon potential. Initially, the Soviet Union resisted the proposal because, among other reasons, it included a provision enabling personnel assigned to the technical secretariat of the proposed consultative committee to conduct on-site inspections, an essential part of any meaningful and verifiable chemical treaty. Subsequently, however, the Soviets agreed to on-site inspections in the 1986 Stockholm Declaration which followed the Conference on Confidence- and Security-building Measures in Europe. The USSR then agreed in Geneva that on-site inspections are necessary to verify a chemical weapons treaty. Currently, negotiations on the detailed measures required to implement this agreement in principle and to verify compliance with the draft treaty are being pursued in the multilateral CD and in associated bilateral discussions. Although the monitoring provisions of the 1987 Treaty on the Elimination of Intermediate-range and Shorter-range Missiles are quite different—and far less demanding—than those required for a chemical weapon convention, the progress in monitoring accomplished in the INF treaty suggests that successful negotiation of a chemical weapon agreement is possible.

In chapter four, "Technical Means of Verifying Chemical Weapons Arms Control Agreements," Franklin Walker, a senior scientist in the treaty verification program of the Lawrence Livermore National Laboratory, analyzes the potential applications of new technologies in light of the demanding requirements for effective inspection and monitoring systems. Not only must the instruments for on-site inspection be highly accurate and capable of identifying a wide range of lethal agents, but they must be portable, rugged, and self-contained. Inspections would have to be conducted swiftly, with limited advance notice, and possibly in relatively inaccessible parts of the globe. The optimistic prognosis for these technologies supports the idea that a concerted research and development program should be initiated to bring them to fruition. The needed instruments must be able to corroborate or disprove suspected violations of the treaty by accurately and quickly identifying prohibited chemical agents. The wide variety of chemicals with weapons potential, including known and unknown substances, means that several of the technologies capable of detecting and identifying prohibited chemicals would need to be used in tandem.

The routine and continued analysis of chemical stocks by remote sensors at chemical production, storage, transportation, and destruction facilities would be a second invaluable tool to verify treaty compliance. Many of these monitors would be versions of automated control systems now used for commercial purposes throughout the industrial world. Others would be instruments, to be used by inspectors, that are modified for unmanned, remote performance. The monitors would provide the information on which the need for challenge inspections would be based and could possibly confirm noncompliance themselves. The data collected could go directly to the facilities of a standing Treaty Compliance Committee and might be capable of providing immediate warning of treaty violations. Such uses of remote, unmanned sensors could greatly reduce requirements for human inspections, thus facilitating verification of the treaty. On March 7, 1987, the executive committee issued a press release supporting the further development of technologies necessary for the verification of chemical weapons treaties and published Dr. Walker's paper under a separate cover. Subsequently, the FY 1988 Defense Authorization Bill was amended by the Congress to authorize funding for a three-year research and development program proposed by the U.S. Army Chemical Research, Development & Engineering Center for the purpose of developing and demonstrating the technical means of verifying the proposed convention. This program could also benefit U.S. negotiators as they engage in the difficult task of reaching agreement on exact treaty provisions.

During the committee's discussions of verification and monitoring, it became apparent that it is critical for the U.S. government to be well organized to address these issues. Apparently, no single policy organization in the government assumes responsibility either for developing technologies to facilitate arms control verification or for assessing the verifiability of agreements prior to their submission to negotiation; the mishandling of the U.S.-proposed chemical weapons convention illustrates the difficulties which result from inadequate policy coordination. The question of verification responsibility within the government is not new. Where responsibility has been ambiguous or delegated to insufficiently powerful agencies within the bureaucracy, progress on an issue has been slow or nonexistent. Different presidents have experimented with a variety of organizational norms for arms control verification. To assess the current status of this important issue, the committee commissioned a paper from Michael Krepon of the Carnegie Endowment for International Peace on "U.S. Government Organization for Arms Control Verification and Compliance" (chapter five).

Mr. Krepon evaluates the bureaucratic organization for assessing the verification of arms control agreements within the executive branch from Richard Nixon to Ronald Reagan. He argues that the issue of arms control verification was addressed most effectively by those administrations with centralized control over verification assessments. As a result, he recommends a strong National Security Council (NSC) role in the assessment process. Mr. Krepon further recommends the establishment of annual interagency reviews of verification techniques and suggests that the Arms Control and Disarmament Agency (ACDA) play a larger role in verification research, but have less responsibility for establishing verification policy.

The committee also discussed new problems that might result from the superpowers' use of space. The expanding military roles of spacecraft have exacerbated existing concerns over the security of national space assets, including communications, weather, and reconnaissance satellites. As a result, the need for formal international guidelines for the conduct of space operations has become increasingly apparent.

As Paul Stares of The Brookings Institution explains in "Rules of the Road for Space Operations" (chapter six), the successful negotiation of an agreement on space operations could serve to reduce the risk of misunderstandings or crises resulting from unexplained incidents; to reduce the risk of surprise attack; and, to set a precedent for future cooperation on space issues as nations further explore its uses for scientific, commercial and military purposes. Several types of provisions could form the basis of such an agreement. Alternatives presented include a prohibition on the use of force in space, an "incidents in space" agreement, establishment of "keep-out" self-defense zones for space assets, crisis management procedures, and cooperative monitoring and collateral measures.

Guidelines for space operations would be most useful for the superpowers since they are actively engaged in space operations. The possible precedential value, however, is also important. The establishment today by the United States and the Soviet Union of cooperative procedures for space operations could reduce future aggravation of the problem as a result of the increasing use of space by third countries.

Peacekeeping operations provide yet another area where technological innovation can be exploited to promote international stability. Efforts to prevent the outbreak of, or to terminate, hostilities between states in the Third World have become increasingly important. Not only do Third World conflicts cause great destruction in themselves, but each also contains the seeds of more far-ranging conflicts between the world's advanced military powers and could even potentially catalyze a nuclear conflict. Technology can be used in support of peacekeeping operations by helping to maintain cease-fires and improving the prospects for negotiated settlements. Remote-sensing devices, for example, emplaced so as to monitor the troop movements of both Israel and Egypt have contributed significantly to the ability of U.S. peacekeeping forces in the Sinai to help maintain the Camp David accords. As discussed by William Lewis of the Institute for Sino-Soviet Studies of George Washington University, in chapter seven, similar applications are possible in other world trouble spots.

Measures that can be taken to contribute to peacekeeping efforts include the use of sensors to detect hostile troop movements on either side of a demilitarized neutral zone between two enemy nations and to prevent border crossings by unwelcome forces. In "Technology and International Peacekeeping Forces," Dr. Lewis identifies a discrepancy between the capabilities and responsibilities of peacekeeping forces and reviews a number of technologies that could enhance the capabilities of these forces. Suggested technical means to facilitate peacekeeping operations are the provision of comprehensive surveillance capabilities to enable peacekeeping missions to separate and monitor hostile forces and to observe international shipping lanes. Lewis concludes with a review of the major aspects of this issue which he believes should be addressed by the U.S. government, international agencies, and others interested in the future prospects for peacekeeping missions.

Nuclear weapons pose special dangers because their existence opens up the possibility of accidental or unauthorized use. Any such use would be catastrophic not only because of the resulting destruction and environmental damage, but also because it would incur the danger of a response by another nuclear power. The most feared scenario is that of a full-scale nuclear exchange between the United States and the Soviet Union as a result of an accident or unauthorized use.

From the advent of the nuclear era, the United States has taken many precautions to prevent the accidental or unauthorized use of nuclear weapons. Technical devices to prevent unintentional detonations were inserted into even the earliest nuclear bombs. More sophisticated devices have been installed as technological advances have made them available. It is widely believed that the Soviets have taken similar precautions. If so, it is fair to say that the arsenals of the superpowers are as safe as they ever have been.

In "Safety, Security, and Control of Nuclear Weapons," (chapter eight), Gerald Johnson provides a historical overview of efforts to decrease the likelihood of unauthorized or accidental detonation of nuclear weapons and to prevent a nuclear yield in the event of an accidental detonation of the conventional explosive contained in nuclear weapons. Noting the success of these efforts, Johnson identifies the growing threat of organized terrorism as cause for additional efforts. He recommends several steps to further improve the security of nuclear weapons, including establishment of a Defense Department committee to monitor information on the location of nuclear weapons and the installation of a device to disable nuclear warheads if an attempt were made to bypass other protective systems.

Johnson identifies the arsenals of such emerging nuclear states as India, Israel, Pakistan, and South Africa as the weakest links in the security of nuclear weapons. Many of these countries lack the technology, and in some cases may be unwilling to devote the resources, to install permissive action links (PALs) and other sophisticated devices to protect their nuclear weapons. Johnson addresses the possibility of a U.S. unilateral or bilateral initiative to promote the use of protective technologies by these countries. This problem is especially sensitive because the United States does not want to appear to be promoting nuclear proliferation or providing information that could help those states to design nuclear weapons.

Unprotected nuclear weapons threaten many nations, not only those that possess them. Given the devastation that could result from an accidental detonation and the increased risk that such an episode could result in an inadvertent nuclear war, it is essential that all nuclear states protect their arsenals with these devices. Ways to share technologies that have proven effective at protecting U.S. nuclear weapons without compromising weapons designs or nonproliferation objectives deserve serious consideration.

The committee discussed the possibility of a U.S. government publication on the design of devices to improve the security of nuclear weapons. The potential benefit of such a publication would be that nations with unprotected nuclear arsenals would not have to undertake costly research programs, which many may not be inclined or qualified to do to increase the security of their arsenals. Protecting nuclear arsenals in any nation from unauthorized uses is very much in the interest of the United States. The PAL designs provided would necessarily be rudimentary, but could be effective nonetheless. Only designs for rudimentary devices that are no longer a part of the U.S. arsenal would be provided to avoid revealing sensitive information about U.S. protective systems currently being used in the United States.

Issuance of a government publication is preferable to providing information directly to potential proliferators, as it would avoid the impression that the United States favored that nation's advancement of a nuclear weapons capability. Still, one concern about a publication on PAL designs is that any official dissemination of information relating to nuclear weapons may appear to encourage nuclear proliferation. Such a publication would have to omit all information useful to the construction of nuclear weapons themselves in order not to encourage—or appear to encourage—nuclear proliferation.

The committee took no position on the desirability of releasing PAL design information.

In chapter nine, "Reducing the Risk of Nuclear War with Permissive Action Links," Dan Caldwell of Pepperdine University and Peter Zimmerman of the Carnegie Endowment discuss the use of safety and security devices for nuclear weapons deployed on U.S. naval vessels. Currently, no permissive action links exist on nuclear weapons deployed aboard U.S. Navy ships and submarines. The communication problems intrinsic to submarine operations preclude the installation of PALs on their nuclear missiles. Naval commanders of both submarines and surface ships are also accustomed to wide latitude in their operations, especially because flexibility is one of the most important assets the navy offers. Moreover, PALs were originally introduced for use on weapons stationed on foreign soil—to preclude their use by non-U.S. nationals—and U.S. naval vessels are considered U.S. territory.

Caldwell and Zimmerman discuss the reasons why PALs should be installed on U.S. surface ships, counterarguments notwithstanding. Unlike submarines, surface ships do not have the same communication problems that prevent transmission of the codes to release the PALs and thereby make operable the nuclear weapons in which they are installed. The codes could accompany the authorization to use nuclear weapons required for the ship's captains and other officers to release the weapons in any event.

Whereas PALs were intended originally only for use on foreign soil, it has since been recognized that they contribute to the security of all nuclear weapons. Ships, moreover, sometimes enter foreign ports, which may significantly increase the risk of theft or unauthorized use of nuclear weapons stationed on board. Drs. Caldwell and Zimmerman conclude that while it would not be appropriate to install PALs on submarines, their installation on the nuclear weapons of surface ships would considerably enhance the security of this portion of the U.S. arsenal without sacrificing the element of flexibility important to the navy.

The United States has long regarded technological prowess as one of its greatest endowments. The use of technology to enhance international security and stability offers the opportunity to apply the fruits of progress to contemporary problems of great importance. Improving the means to support political agreements whose purpose is to limit conflict between nations and to reduce the likelihood of war is a necessary step toward lessening the dangers inherent in the existence of large numbers of weapons of mass destruction throughout the world. As the essays in this volume demonstrate, a variety of innovations offer the potential to reduce international conflict and to reaffirm the ability of the United States to channel technological change in positive directions.

CONFIDENCE BUILDING WITH UNMANNED SENSORS IN CENTRAL EUROPE

BY VICTOR UTGOFF AND IVAN OELRICH

Large armies have confronted each other in central Europe for forty years now. No historical precedent exists for such enormous concentrations of military force, standing at such high states of readiness, facing one another separated by just a few kilometers, yet remaining at peace. Despite the forty years of peace since World War II, however, both sides remain concerned about the dangers and costs posed by these forces.

The direct approach to reducing the costs and risks of this continuing confrontation has been tried. The long-running Mutual and Balanced Force Reduction (MBFR) talks have attempted to reach agreement on reduced force levels, primarily manpower levels, on both sides. They have failed thus far because, among other reasons, the two sides have not been able to deal with key asymmetries in their forces.

Another approach is to accept for now the size of the forces and to work on ways to reduce the likelihood of war, especially a war that results from a misunderstanding or a poorly controlled spiral of escalation. This approach seeks to reduce the uncertainty concerning the intentions of the other side and also to reduce the opportunity for surprise. Efforts to these ends are called "confidence-building measures" (CBMs) and are the subject of ongoing negotiations.

The 1986 Stockholm Agreement signed by the United States, Canada, and thirty-three European nations, including all of the members of both military alliances, requires prior notification of all maneuvers exceeding a certain size. This is a good example of a CBM—that is, a circumstance in which a nation may have no hostile intentions and may want a way to assure its neighbors of that. There are other instances when nations may wish to reassure others of their lack of hostile intent, and other mechanisms have been developed to make this possible. The U.S.-Soviet hot line is one example; the recently established nuclear risk reduction centers are another.

Installing Unmanned Sensors as a Confidence-Building Measure

In this chapter, the results are reported of a study to determine whether cooperatively-placed, unmanned sensors could provide trustworthy

13

reassurance that preparations for a conventional attack in Europe are not taking place. The study concludes that such a system of sensors is feasible; indeed, more than one approach is possible. One arrangement of sensors and monitors that would be effective as a CBM is presented and a few variations are mentioned. Although central Europe is the area of study, the results could be applied in other regions as well.

Briefly, the complete system would consist of cooperatively-placed, unmanned, short-range seismic sensors arrayed around armored vehicle garrisons, plus lines of sensors that would detect the movement of armored forces to the front from outside the general area where the garrison monitors are located. The garrison sensors—which are little more than buried microphones—would be able to detect the movement of armored vehicles within the garrisons as well as the movement of trucks and armored vehicles into and out of the garrisons. The lines of sensors would consist of television cameras at the major roads and railroads, seismic sensors to fill the gaps between the roads, and passive sonars where rivers make up part of the line.

The proposed sensors do not require any sophisticated technology, indeed, sensors similar to those needed are already commercially available. The data from the sensors would be available to the monitored side directly and transmitted to the other side for analysis and evaluation, preferably via satellite. The system is depicted schematically in figure 1.

Armored forces are attractive objects to monitor for three reasons: (1) they are essential for a successful conventional attack; (2) they have characteristics sufficiently different from normal civilian vehicles to make them relatively easy to identify; and (3) the time required to assemble and prepare large formations of armored vehicles for battle is long enough to give useful amounts of warning.

The primary value of the monitoring system would be the continuous signal confirming that preparations for an attack were not under way. Concentrating forces is key to a successful conventional attack, and the time required to prepare forces for an attack, concentrate in attack formation, and move forward is greater than the time required for the intended victim to disperse its forces into defensive positions. Attackers traditionally have used secrecy and deception to compensate for this relative disadvantage. If the proposed monitoring system made each side confident that it would learn at an early stage of the other's preparations for large movements of armor, then each side could afford to delay some of its most provocative final defensive counteractions, such as deploying its own armored forces out of garrison.

The monitoring system would do nothing to make war impossible, and one should expect it to be turned off if either side were actually planning an attack, but turning off the system would itself provide an important indication of hostile intent.

Figure 1
Using Unmanned Sensors to Monitor Force Concentrations

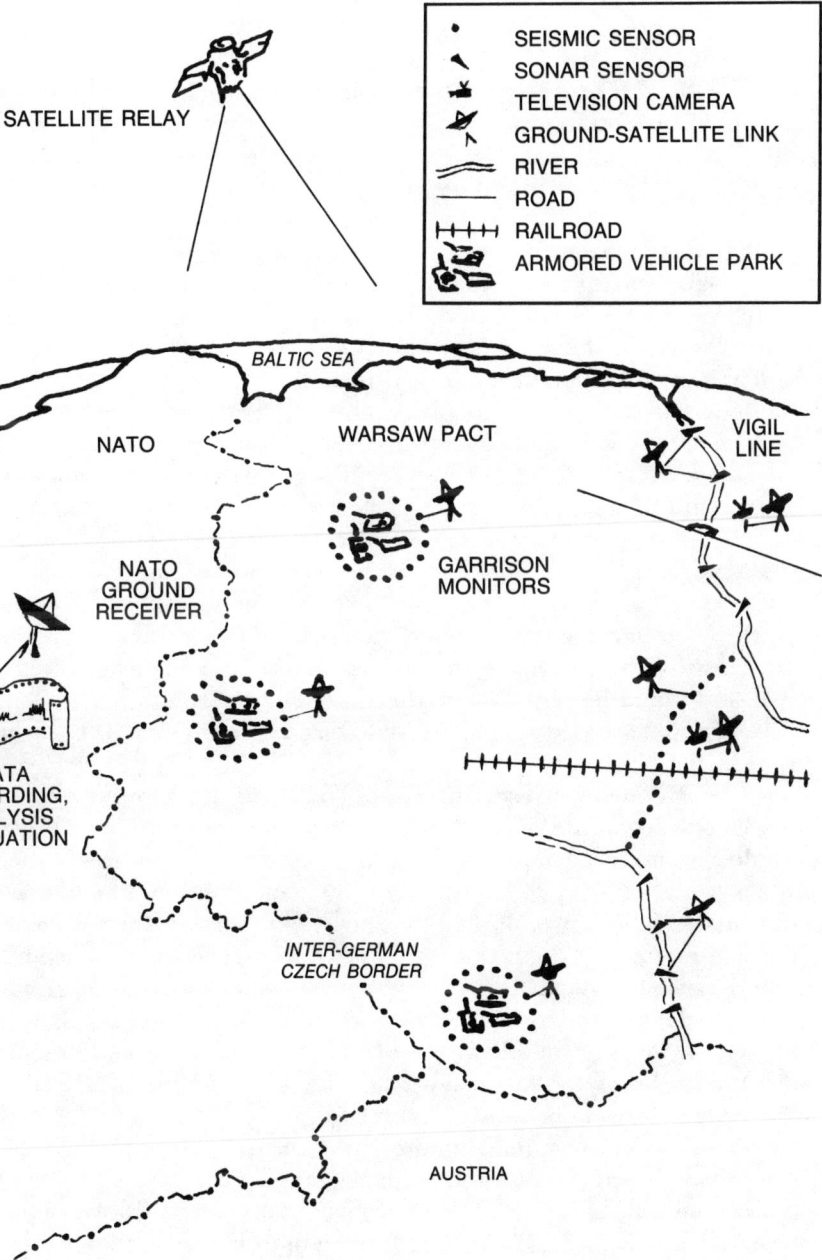

The system could also be used to help monitor future arms reductions or redeployments designed to enhance the stability of the conventional force balance. And over time, the system would show a clearer picture of normal activities. This picture would provide a better background against which to assess whether suspect Warsaw Pact activities were preparations for an actual attack.

Finally, the process of building the system could itself build confidence. The negotiation would include extensive data exchanges on force dispositions, additional explanations of maneuvers, and inevitably some exchanges of views of military matters.

A Phased Approach to Creating a Monitoring System

The complete monitoring system would be extensive, probably rather intrusive, and unlike any arrangement that the two blocs have in place now. Therefore, the negotiation and establishment of the complete system in one step might be an unrealistic objective. Moreover, establishing the system in several steps might be more efficient and effective, because the two sides could learn from early steps how to implement better the later stages. It may be best, therefore, to negotiate the overall goals of the system and its general design and then to implement it in stages.

As a first step, the Warsaw Pact and NATO could agree to monitor some small number of each other's garrisons, perhaps just one or two garrisons on either side. NATO would provide sensors for monitoring the Warsaw Pact garrisons and, under Warsaw Pact observation, would install the sensors. The data from the sensors would be made available to the Pact and would be sent back to the NATO side for analysis. The Warsaw Pact, of course, would similarly install sensors around one or two agreed sites in the NATO area.

Clearly, one or two garrisons could not spell the difference between the success or failure of an attack, so this limited version of the system would give no information that could be used to determine whether an attack were planned. The purpose of this stage would not be to provide information about overall military activity but to give information about the sensors themselves: how they should be cooperatively installed; how well they would operate; how reliable they would be; what information could be derived from the data they provide; and whether the data from the sensors might yield other types of information (for example, on movement of nuclear weapons) that neither side may want to make available.

If both sides were satisfied with this initial deployment, they could move to a more widespread monitoring of the garrisons of the two blocs. The sensor arrangement for individual garrisons that proved effective in the first step would be replicated at many more garrisons. What would be tested in this phase is how the information from many monitoring sites would be handled collectively. This phase would provide experience with transmitting, receiving, and analyzing large quantities of data, integrating and comparing data from a variety of sources, and ultimately synthesizing an estimate of the level of armored forces' activity within the collection of monitored garrisons.

Once both sides were satisfied that a more extensive monitoring system was workable, the system could be expanded again by including more of the garrisons in the monitored area, by monitoring garrisons in a much larger geographic area, or by constructing a string of sensors to give information about the movement of forces from unmonitored areas into the monitored areas. Some combination of these measures could be the next step: a string of sensors could detect general movement into the monitored area, and particularly important groups of forces—for example, transportation units with tank transporters or engineering units with bridging equipment—could be monitored even though they were located on the far side of the sensor string.

In a final phase, the sensor and monitoring system could be combined with arms reductions or redeployments. Unmanned sensors, for example, could monitor a reduction in the armored forces close to the inter-bloc border. They also could be used to monitor an agreement to separate forces into subunits, a process that would delay even further preparations for large-scale offensive operations. Other forces in addition to armored forces—for example, tactical aircraft—also could be monitored. In general, as the monitoring system became more extensive, it could verify more extensive types of reductions. If the system eventually were extended over all of Europe, from the Atlantic to the Urals, then major reductions in forces might be monitored.

The Components and Functioning of the Monitoring System

The individual devices that receive information from their environment are called "sensors"; a set of sensors arrayed around, say, a garrison site is called a "monitor"; and all the monitors and sensors together are called the monitoring "system."

The final monitoring system would consist of three basic components: (1) monitors of armored vehicle garrisons to detect activity at a particular point, (2) strings of sensors to detect movement of armored vehicles from the areas that remain unmonitored into the broad areas that are monitored, and (3) a means of transmitting the data from all of the sensors and receiving them on the other side of the inter-bloc border.

Garrison Monitors. The garrison monitors would consist of a set of seismic sensors, or buried microphones, arrayed around the garrison site. The set of sensors around each garrison would be connected to a small satellite ground station that would transmit the data to a satellite for relay across the inter-bloc border. Just one or two sensors could detect armored

17

vehicle movement into and out of the garrison, but better monitoring is possible if the sensors are spaced every hundred meters or so around the garrison. Even though this close spacing may require as many as thirty sensors for a large garrison site, it is feasible because the sensors are small and fairly inexpensive. With more closely spaced sensors, the monitor could detect trucks going into the site as well as armored vehicles moving within. Detecting truck activity could provide additional information about preparation of forces for movement. The encircling sensors would provide no barrier to the movement of monitored forces; in fact, the sensors and connecting cables could be completely buried. In this case, the only part of the monitor that would be visible would be the antenna that relays the data out.

Sensor String Vigil Line. At later stages in the development of the system, increasing numbers of garrison sites would be monitored. If all the garrisons in Eastern Europe, from the German Democratic Republic to far back into the Soviet Union, were monitored, then garrison monitors alone could provide confidence to NATO that final preparations for an attack were not taking place. If the garrisons were not monitored in such great depth, however, then large forces might be left unmonitored in the western Soviet Union. These unmonitored forces could move undetected into and through the monitored area by passing between the monitored garrison sites. Preventing this would require a string of sensors to detect movement of armored forces into the monitored area.

As with the garrison monitors, the sensor string separating the monitored from unmonitored areas would provide no actual physical barrier to movement, but would only sense the passage of armored vehicles. This "vigil line" could be made up of several different types of sensors. The backbone of any sensor line would probably be seismic sensors similar to those used to surround the monitored garrisons. Where significant roads crossed the vigil line, however, visual or infrared cameras would be required to supplement the seismic sensors because the sensors could not distinguish between normal civilian truck traffic on the roads and military trucks carrying armored vehicles. Making rivers part of the vigil line and monitoring bridges with cameras would be an attractive means of construction of the sensor line. Because most Warsaw Pact armored vehicles have some capability to ford rivers and could bypass the bridges, it may also be necessary to monitor the rivers themselves, at least the fordable stretches of the rivers. Calculations have been made that show that rivers are excellent sound channels and that acoustic sensors placed in the rivers should easily detect swimming or fording armored vehicles.

The placement of the vigil string would be somewhat arbitrary as long as it was not located too close to the border. If it were too close, then the system could be defeated simply by massing forces on the far side and driving through it. The vigil string might work perfectly, but the warning time it would be able to provide would be too short for the defending side to react effectively. Keeping in mind the advantage of using rivers and the long detection ranges that can be achieved in them with acoustic sensors, one can conclude that the closest vigil line on the Warsaw Pact side that would still be useful should follow the Oder River in the north from the Baltic Sea to Krystkowia, Poland. It should then run roughly in a straight line overland to Decin, Czechoslovakia, where the line would pick up the Elbe River and its tributary, the Vltava. A final short overland segment would connect to the Austrian border. This course, shown in figure 2, requires about 200 kilometers (km) of seismic sensor strings running cross-country and about 620 km of sonar sensor strings running along navigable rivers. A vigil line running this course would keep unmonitored forces an average of 150 km from the inter-German border, somewhat less in the south and somewhat more in the better tank terrain of the north. The comparable system that the Warsaw Pact would place in NATO territory would probably follow the Rhine.

The Soviets could try to defeat a vigil line along the Oder by massing forces just east of the line and then sprinting through to the inter-bloc border. The distance is about 200 km in the north, however, and moving armored divisions these distances takes a surprisingly long time because of the importance of keeping units together and maintaining organization among them. Nonetheless, if more warning time were considered essential, the vigil line could be placed farther back, perhaps along the Vistula River, or along the Polish-Soviet border, or even inside the Soviet Union.

Data Transmission. The data from the sensors must be sent across the inter-bloc border. During the first stage, when only one or two garrisons on either side were monitored in order to learn more about the utility of the sensors, the data could be sent over commercial telephone lines. Later, as the quantities of data increase and the data becomes more useful and more important, the monitoring side may want to transmit the data via satellite for greater convenience and reliability of transmission.

Sensors and Other Technical Components

A detailed technical explanation of the sensors that could be used for the monitoring system is outside the scope of this chapter. That the technology is available and that the monitoring system should be affordable is all that can be shown here. What follows are brief descriptions of existing similar sensors and how they would work in the suggested confidence-building monitoring system.

Seismic Sensors. Seismic sensors for the detection of vehicles have been developed for battlefield reconnaissance—for example, the Remotely Monitored Battlefield Sensor System (REMBASS) currently deployed with the U.S.

Figure 2
Illustrated Sensor Line

DENMARK

HAMBURG

GDANS

SZCZECIN

POLAND

BERLIN

← Oder River

AMSTERDAM   W. GERMANY

NETHERLANDS

E. GERMANY

← Krystkowia

WROCLOW

BELGIUM   FRANKFURT

← Decin

← Elbe River

PRAGUE

LUXEMBOURG

CZECHOSLOVAKIA

← Vltava River

FRANCE

VIENNA

MUNICH

AUSTRIA

ZURICH

SWITZERLAND

COUNTRIES
CITIES
International Boundaries ——
Barrier Line •••••

Army. These fully automated sensors can detect tracked vehicles at a range of 350 meters.[1] The REMBASS seismic pickup is a small microphone buried just under the ground. Practical use on the battlefield requires that the entire system be small and light (the latest version fits into the pocket of battle fatigue trousers), and the small microphone thus required inevitably has less than ideal acoustic coupling to the surrounding soil. A permanently installed sensor, however, could have a more sensitive pickup attached to a well-buried block of material specifically designed for good acoustic coupling with the surrounding soil. With better acoustic coupling, the seismic detector would be even more sensitive.

Because careful permanent emplacement of seismic sensors is impractical for battlefield applications, measurements have not been made of how well REMBASS sensors thus installed might work. Nevertheless, industry representatives have estimated informally that a 60 percent increase in sensitivity would be easy to achieve. The baseline costs shown below do not assume any improvement over currently demonstrated REMBASS capabilities. Indeed, in the cases discussed here, conservative estimates are made of the sensitivity and resultant number of sensors required.

Cameras. Seismic sensors may have the greatest difficulty in detecting military vehicles that stay only on the roads. Specifically, if armored vehicles are transported on large trucks, then the sensors will hear the trucks, but military truck traffic could be hidden in normal civilian truck traffic. We suggest, then, that major roads, and also railroads, be monitored with simple optical or infrared cameras. Monitoring roads and railroads where they bridge rivers would be particularly useful.

Achieving good camera resolution is easy—commercial television cameras have between 400 to 800 lines per frame, far better resolution than required here—but the amount of data generated with high resolution would be expensive to transmit. One way to use the high resolution of the camera without taxing limited data transmission capabilities is to transmit only that part of the image that changes from frame to frame. In this way a fairly wide angle lens could observe a broad view of the monitored area, and pointing the camera is unnecessary.

A section of the picture containing only 32×32 points, each with any of eight shades of grey, would allow the visual identification of tracked vehicles. (Armored vehicles are too large to be transported out of sight inside normal enclosed rail cars and trucks.) A 32×32 subsection of the view would contain 3,000 data bits. If the data transmission rate were limited to 3,000 bits a second (the capability of a telephone line), one subsection image could be sent every second. One image every second is not sufficient to report every vehicle on busy major roads, but an adequate

---

[1]Joseph Santarelli, GE Automated Systems, Burlington, Massachusetts, personal communication.

sample could be taken. The best sampling system would combine random camera activation with activation by a seismic sensor upon detection of a heavy vehicle. The random activations would help to defeat attempts to tamper with the cameras, as will be discussed later.

Riverine Sonar. Most Soviet combat vehicles can either float and propel themselves through the water or, with specially attached snorkels, they can drive across river bottoms. Thus, in principle, the vehicles could go around bridge monitors just as they could avoid road monitors by traveling cross-country. Analogous to the use of seismic sensors to detect off-road traffic, underwater microphones could detect the sound of "swimming" or "snorkeling" armored vehicles. Substantial information on marine sonar and acoustics is available, but, not surprisingly, no information seems to be available on riverine sonars except those used by fishermen to find schools of fish. Therefore, here are the authors' calculations of sound transmission in rivers using standard ray-tracing techniques. Assuming total radiated acoustic energy of 1 watt from a swimming or snorkeling vehicle (torpedoes generate 10–20 watts), a straight river with semicircular cross-section, and muddy bottom reflection losses (the worst case), one can calculate that swimming vehicles should be easily detectable several kilometers away from an acoustic sensor. The only assumption in the calculation that is not conservative is that of a straight river. In the cost estimates below, that problem is dealt with by assuming at least one sensor per straight section of river regardless of how short its length.

Data Links. Because of the large number of sensors, the transmission of data would be a substantial task. One possibility would be to use the existing telephone system, but that would have some disadvantages, the foremost being the lack of control that the monitoring side would have over the quality and reliability of the transmission.

Although perhaps more expensive, the data might therefore be better sent out via satellite transponder. Satellite ground links also would be costly, but if several sensors were closely spaced, as they would be around a garrison area, then the sensors could be linked together by cable and share a single satellite ground station.

Data Management and Authentication

The amount of data transmitted is an important determinant of the complexity and cost of the data links. Sending more data than required would incur unnecessary costs, but sending too little data would run too great a risk of building a system that would provide no real confidence. Data rate requirements could be reduced substantially if some on-site data processing were accomplished. For example, in the extreme, a computer at the site could be given a program that analyzes the seismic data and gives a simple yes or no answer to the question, "Any tanks?"—this is how the REMBASS system now operates. The problem with this approach is that unforeseen and perhaps simple changes in the vehicles' signature might confuse the program, and such a failure might go undetected for extended periods. Also, safe encryption of so simple a signal could be difficult. An intermediate level of data transmission requirements would occur if the data were aggregated to some extent–for example, different types of vehicles generate noise in different frequency bands, so rather than transmitting the entire seismometer output, transmitting only the ratios of intensities at a few select frequencies would allow identification of most vehicles by type most of the time.

Sending raw data—the direct seismometer readings, for example—would require that an extremely large amount of data be transmitted. It would give maximum flexibility to analysts on the receiving side, however, and would yield maximum information. Avoiding detection by altering the signatures of monitored vehicles—for example, trying to alter the seismic signal of a moving tank—would be extremely difficult. Even if this were attempted, and the analyst at the other end did not know what the new sound was, with raw data he would know that something unusual was happening.

Periodic sampling of raw data seems to offer the best compromise between high-quality data and reasonable data transmission requirements. Randomly chosen twenty-second samples of the full data every three to five minutes would detect significant armor movements, making the system difficult to fool while reducing the data transmission requirement by a factor of ten. Sampling is particularly attractive with a two-way data link. The monitoring party could then instruct the monitors to send more data from areas of suspicious activity.

Any unmanned on-site monitoring scheme could be defeated if it were possible to block the actual data transmission from the sensors and replace it with bogus data falsely indicating that all were well. Fortunately, one can guarantee detection of substituted data by encrypting the data with a code that the monitored country cannot mimic. Once the data are encrypted, they could be transmitted over any open lines. Substituting a bogus transmission may be physically possible, but any substitution would be immediately obvious upon decryption by the monitoring party.

If the data were encrypted at the sensor site before transmission, the monitored country might fear that additional data could be sent out beyond that strictly required for the agreed monitoring system. This objection could be met, in principle, by using what are called public key encryption systems.[2] Such systems allow one party with an encryption key to encrypt data and anyone with the decryption key to decrypt them. By

---

[2]Martin E. Hellman, "The Mathematics of Public-Key Cryptography," *Scientific American*, vol. 241, no. 2 (August 1979):146.

this means, the monitoring party could encrypt to ensure data authenticity while the monitored party would have immediate access to the data sent out. In practice, the problem of "extra" information being sent out is solved by making certain that the sensors are incapable of picking up any more information than that to which they are entitled. For example, each side might worry that the other's sensors contain hidden radiation detectors that could give information about the movement of nuclear weapons. Therefore, each side would probably want to dismantle and inspect some randomly selected fraction of the other's sensors to make certain that they can collect only agreed types of data.

Tampering and Tamper Detection

The problem of "tamperproofing" the sensors—ensuring that they cannot be physically altered in such a way as to send out false data—is similar to the problem of data authentication. Constructing a sensor that cannot be tampered with while operating for long periods unattended in another country is almost certainly impossible. But such security is not necessary. Making certain that tampering will always be detected quickly is good enough. The system would not require tamper *proofing*, only tamper *detection*. Like data authentication, this problem has been worked out in great detail for the unmanned seismic stations that would monitor a comprehensive nuclear test ban treaty.

Among the important attributes of the suggested sensors—cameras, sonar, and seismic detectors—is the ability to monitor themselves. Seismic detectors are sensitive buried microphones. Any attempt to work around them or even approach them would be detected. The associated vibrations can be picked up from the ground directly through that part of the sensor housing buried firmly in the soil. The sensor need not necessarily have any microphone or other detector external to it. It could be, literally, a block of concrete with a signal wire coming out. The only way to tamper with it would be to dig it up, which would be easily heard.

The cameras at roads or railroads could be mounted in pairs on either side of the roadway and so positioned that each could view the other to detect tampering. Or, the cameras and seismic sensors could be located within the same protective envelope so that the seismic sensor could be used to detect tampering. A single camera would then be sufficient. One could also try to defeat the sensors by blocking or altering the path carrying the signal into it. For example, a simple infrared sensor could be defeated by placing a shroud over it that blocked the heat emitted by passing armored vehicles. For seismic sensors, the information path is the earth. Any attempt to isolate the sensor from the surrounding earth by digging around the sensor station would be detected by the seismometer itself. Similarly, blocking the view of the cameras, even with backdrops, would be detected by the cameras.

Theoretically, there could be a problem with a purely passive riverine sonar. It may be possible, although no doubt difficult, silently to place a soundproof box over the sonar and block the signal. Although this would also block the background noise and almost certainly be detectable, for extra confidence this potential attack on the sonar could be countered by making the sonar active. If the sonar occasionally chirped, the operator at the other end would learn the characteristics of the local echo. Reproducing the echo, in frequency response, direction, and time delay along with the normal flow and background noise of the river would be impossible for the soundproof box. The sonar would thus be a bit more complex than the seismic sensor. More than one sound pickup may be required, the sonar may need to be active, and it would be more difficult to install and service. However, because perhaps as few as one-twentieth as many sonar sensors are required as seismic sensors, exploiting the natural river barriers still seems likely to be worthwhile.

Besides blocking the signal into the sensors, the monitored party could block the signal coming out of the sensors: if the system sent its data out over telephone lines, the monitored party could cause a failure in the telephone system. Although an arranged failure would deprive the monitoring side of the system's data, the fact of the failure is obvious and should not deceive anyone.

The sensors also could be defeated by raising the noise level of the environment. Once the monitors were in place, for example, the monitored country could begin extensive, and seemingly never-ending, earth-moving work near important sensor sites. Treaty provisions may require some specific reference to this problem and how to relocate monitors when background noise increases for some legitimate reason. Note, though, that both of these forms of "tampering"—killing the transmission links and creating nuisance noise—are really just harassment. If they occurred during a crisis, they would constitute warning themselves. If they occurred repeatedly, the monitoring party would have to question its involvement in cooperative confidence-building activities.

Achieving reliable tamper detection is critical to the success of the proposed system. For the monitoring system to have any value at all, both sides must have essentially perfect confidence that it is reliable and not being used by the other side to deceive them. To be certain of the security of the system, there probably should be a continuing "red team" effort—that is, a group of skilled engineers and military personnel constantly trying to find weak points.

Power

Another important consideration is the power requirements of the monitoring system. The seismic sensors would require a watt or less. The cameras would require on the order of 10 watts, and, if cooling were required for an infrared camera, the cooler would require 100 watts or more.[3] If the camera were not infrared or low light sensitive, it would require some form of external lighting, like a street light, which could use a kilowatt of power. The power requirements of any calculation or encryption device would be negligible.

The power requirement of the data transmission device would vary closely with the required data rate, and, as was pointed out above, that rate is somewhat flexible. Commercial satellite links that carry 56 thousand bits per second use 800 watts;[4] this figure can be taken as an upper bound of the power required by the satellite link for a garrison monitor site. Gross power required by the transmitter would be roughly proportional to data rate. Because of the potentially lower data rate requirements of the satellite links in the system described here, a transmitter adequate for the monitor at one garrison site might require as little as 100 watts. The total power requirements per garrison monitor and per kilometer of overland vigil line, then, would be several hundred watts or less.

When considering power sources, the prospects are rather limited. Batteries cannot store enough energy to allow long enough intervals between maintenance visits. Similarly, motor-generators probably do not have adequate reliability. Some fuel cells that are under development could be a possibility for the future. The problem would be made much simpler if the monitor stations had access to commercial line power. Using commercial line power may seem to have potential problems with reliability, but the monitor stations could store enough energy in batteries to operate autonomously for the few days at most that commercial power might get knocked out. Moreover, because the monitor stations would be stationary, they could use large, heavy, and, therefore, cheap batteries. Using sealed, rechargeable, lead-acid batteries, the significant power users, like the satellite ground stations or cooled cameras, would require about $1,000 additional investment cost per required day of operation without line power.[5] For the low power users, like the seismic sensors and built-in encryption computers, the additional cost would be negligible.

Costs of the Monitoring System

Below are some estimates of what the monitoring system might cost. These estimates are based on analogies with existing systems and are only approximations. Existing systems have to meet somewhat different demands than those that would be required of the monitors. Some of these

---

[3]Martin Donabedial, "Cooling Systems," in *The Infrared Handbook*, William Wolfe and George Zissis, eds. (Washington, D.C.: Office of Naval Research, Department of the Navy, 1978).
[4]Robert Courtney, MACOM Government Systems, San Diego, personal communication.
[5]*Handbook of Batteries and Fuel Cells*, David Linden, ed. (New York: McGraw Hill, 1984).

differences would drive monitor costs up relative to the analogous systems. For example, a significant cost might go into ensuring reliable tamper detection, and reliability of the system in general must be high because of the limited number of maintenance visits that might be allowed to the monitoring sites each year.

Other factors would drive the monitor costs down relative to analogous systems, however. For example, the seismic detectors would not have to be light enough to carry nor rugged enough to survive handling by a soldier in the field. Some costs will drop in any case because of technical improvements. The commercially available Very Small Aperture Terminal (VSAT) satellite communications ground terminal now costs about $14,000, but some companies hope to reduce their cost enough in the next few years to sell them to gasoline stations checking credit card information via geosynchronous satellite.[6]

Still other costs are flexible because, as with anything else, "you get what you pay for." For example, a basic fixed video monitor costs $500. Add to it zoom, tilt, and pan controls and enough protection to work outdoors and the cost increases to $1,000. If, however, the video monitor must be able to see in faint light, the camera cost jumps to $5,000.[7] The costs of the seismic sensors, on the other hand, are rather fixed because their important components are already in relatively large-scale production. Specifically, the cost of the seismic monitor would be dominated by the acoustic pickup and associated electronics. These major components in the REMBASS system now cost $7,000.[8]

In summary, the major component costs used here add up to $7,000 for each seismic sensor, $15,000 for each sonar (which is only an active acoustic detector in a more difficult environment), $10,000 for a pair of camera monitors, and $20,000 for a satellite ground station.

Garrison Monitor Costs. To detect preparations for large-scale armored vehicle movements, the sensors required to monitor garrisons must be sensitive enough to detect even slowly moving trucks bringing in maintenance personnel and ammunition. REMBASS has been developed for battlefield applications, so the detection of very slowly moving vehicles has been of little interest, and few relevant measurements have been made. As a very conservative estimate, however, one can assume that even a slowly moving (say, walking speed) truck will generate as much noise as a walking person; the REMBASS system can detect the latter at 50 meters. This means

---

[6]Joseph Aein, RAND Corporation, Santa Monica, California, personal communication.
[7]These cost estimates came from Honeywell and are for normal surveillance cameras, not the still-frame cameras suggested here.
[8]Joseph Santarelli, personal communication.

that sensors need to be placed, at most, every 100 meters around a nominal regimental vehicle park, estimated to be a kilometer in diameter and hence roughly 3,000 meters in circumference.

On average, each garrison thus would require thirty seismic sensors costing $7,000 each, plus one satellite uplink costing $20,000. Each garrison monitor ring thus would cost $230,000. There are approximately 110 Soviet armor or artillery regiments west of the assumed vigil line, so the total cost of acquiring the equipment to monitor this many regiments should be about $25 million. The East Germans and the Czechs have another 40 regiments of armor or artillery. Monitoring them would cost another $10 million. There are certainly other groupings of armored forces smaller than regimental size, but note that the system would not have to monitor every group of vehicles to provide confidence that an attack is not imminent. Both sides may intentionally leave small units unmonitored in order to reduce the overall cost of the system.

Vigil Line Costs. As with the garrison monitors, the vigil line sensors may need to be placed quite close together to guard against any attempt to circumvent them by slowing the tracked vehicles to a crawl to reduce their seismic signal. Therefore, one again assumes a seismic sensor every 100 meters. With such densely spaced sensors, utilizing devices as expensive as the REMBASS sensors, and assuming that ten sensors would share a single satellite uplink, the equipment cost of the vigil line would be $90,000 per kilometer.

Optical sensors are not needed on infrequently used roads. Military truck traffic on these roads would arouse suspicion just because of the increase in traffic volume, and traffic volume would be detectable by the seismic sensors alone. The busier roads that cross the vigil line, however, must be monitored by cameras. If it is decided that just major paved roads and railroads need to be monitored, then only about twenty optical stations would be required if the vigil line followed the suggested route. Monitoring even minor unpaved but all-weather roads would cause the number of required stations to jump to more than a hundred. Even if these roads had to be monitored, the change in total cost would be relatively small. At $30,000 per camera station ($20,000 for the satellite ground station and $10,000 for a pair of cameras), the total cost would still be only $3 million.

If the vigil line makes use of the rivers as described previously, then the 620 km stretch of river could be monitored with 75–80 underwater acoustic detectors, along with 35 additional camera stations to monitor bridges. (It is assumed that all roads that warrant a bridge, warrant an optical sensor.) This requirement would add another $4 million to the cost. Also included is $20 million for the initial installation costs for all three types of sensors.

Total System Costs. The total basic procurement cost, not including installation and maintenance, would thus be $60 million, using conservative estimates of first, the range at which the seismic sensors can detect very slowly moving vehicles; and second, the number of roads that require cameras.

Also estimated are the operations and maintenance costs. Annual material replacement cost, including the sacrifice of some equipment for Warsaw Pact officials to tear down and inspect, might total 10 percent of the procurement cost. This amounts to $6 million per year.

The number of analysts required to monitor the data would be somewhat flexible. A staff of ten is estimated to be on duty at any time for a total staff of 40 on line, including management and administrative personnel. Personnel costs are calculated as though 80 percent of the staff were army master sergeants and 20 percent were captains. Using the same costing procedures that the U.S. Army uses,[9] each team member would cost, on average, $75,000 per person per year for pay, benefits, and retirement accrual. This figure results in a total annual monitoring staff cost of $3 million. The cost of office space and equipment and computer support adds $1 million annually.

With 150 garrisons, 200 cross country seismic sensor groups, 100 camera road monitors, 35 bridge monitors, and 80 sonars, there would be 565 stations to maintain. A team of two people could test and repair, on average, at least one station or sensor group a day. Thus, eight teams on line working five days a week could cover all of the stations every 90 days. The teams would be made up of an army master sergeant and a captain. Using the same costing procedure as for the monitoring staff, each team member's pay, benefits, and retirement would cost an average of $78,000. Because the teams would be traveling out of country, their travel costs may be high, so a total of $150 thousand per person per year is allowed for the repair and inspection personnel. In addition, NATO would have to escort the Warsaw Pact personnel when they are in NATO territory to service their system. Assuming that in-country travel and support costs are lower for escorts, only $100 thousand per person per year is assumed, and one escort per inspection team is assigned. The total cost for inspection, maintenance, and escort personnel would be, then, $3.2 million per year.

The cost of a satellite transponder would be a significant fraction of total operating costs. The number of transponders required depends on the monitor sampling rate. One can assume that the camera monitors would always be available and transmitting at the rate of 3,000 bits per second and that the seismic and acoustic monitors would be sampled approximately one tenth of the time, in which case two transponders would be required for an annual cost of $5 million.

---

[9]All of the military personnel cost data come from the *Military Cost Handbook* (Mountain Valley, California: Data Search Associates, 1986).

All of these costs are summarized in table 1. The initial experiment of monitoring just a couple of garrisons on either side could be started for less than a million dollars in initial procurement and less than a million dollars per year for operations. The complete system proposed here—with 150 regimental garrison monitors, a vigil line running roughly along the German-Polish border, and assuming 50-meter placement of the seismic sensors—would have an initial procurement cost of approximately $60 million, $20 million for installation, and annual operating costs of approximately $18 million. The total ten-year life cycle costs for the full system would be $260 million. If the seismic sensors were sensitive enough to detect even slowly moving tanks at 500 meters, then the ten-year life cycle costs would be less than half, or about $118 million. These costs are low when compared to current intelligence systems that gather comparable quantities of data of comparable quality. The low costs should not be entirely unexpected, however, when one considers that the system, although extensive, would be no more complex than a small town telephone exchange and would be based upon very simple, commercially available technology.

Table 1
Costs of an Unmanned Sensor System
(millions of U.S. dollars)

Procurement and Installation Costs

150 Garrisons at $230,000 each	35
200 km Vigil Line at $90,000/km	18
35 Bridge Camera Stations at $30,000 each	1
80 Sonar Stations at $35,000 each	3
100 Road Camera Stations at $30,000 each	3
Installation	20
TOTAL	$80

Annual Operation and Maintenance

10 percent of Procurement	6
Listening Personnel	3
Office and Equipment	1
Maintenance and Escort	3
Satellite Transponder	5
TOTAL	$18

TOTAL 10-YEAR LIFECYCLE COSTS	$260

Source: Authors' Estimates

Summary and Conclusions

This analysis indicates that a monitoring system that could give reliable warning of surprise attack preparations is feasible and affordable. Because the system would give warning of an attack, any nation actually
planning an attack would almost certainly turn the system off (which, of
course, would give warning itself). But the primary value of the system
would be its continuous emission in normal times of the "non-signal" that
attack preparations were *not* under way. The constant reassurance could
serve as a confidence-building measure in peacetime, help to contain tensions, and avoid unnecessary countermilitary responses during crises.

The monitoring system should be attractive to all the NATO allies. It
could be argued that the system is relatively less attractive to the Warsaw
Pact because they probably have better intelligence on NATO's state of
mobilization than NATO does on their's. Nonetheless, significant political benefits could accrue to both sides. Events during 1987 and 1988 have
demonstrated repeatedly that it is unwise to predict with certainty what
may be possible in the world of arms control.

3

USING TAGS TO MONITOR NUMERICAL LIMITS IN ARMS CONTROL AGREEMENTS

BY STEVEN FETTER AND THOMAS GARWIN

The treaty on intermediate-range nuclear forces (INF) has sanctified the "zero option." It has long been understood that it is easier to verify a complete ban on a weapon system than it is to verify a numerical limit. A complete prohibition is easier to verify because a single sighting of a banned weapon would constitute clear evidence of a violation. Moreover, a complete ban would eliminate training, testing, and repair activities that could serve as a cover for clandestine weapon deployments or could support a sudden breakout from a treaty. Although a total ban may be the best option from the standpoint of verification, this is not realistic for many weapon systems.

In the past, numerical limits could be verified adequately because the weapon systems in question—missile silos, bombers, and ballistic-missile submarines—were hard to conceal from national technical means (NTM) of verification (primarily reconnaissance and electronic intelligence satellites). Unfortunately, changes in technology and in the strategic environment are giving rise to new weapons whose deployment will be difficult to verify using current techniques. Mobile land-based ballistic missiles, for example, are gaining increased prominence in the strategic forces of both sides, primarily because they are less vulnerable to preemptive destruction than immobile silo-based missiles. But mobile missiles are much more difficult to count because they are designed to move around the countryside and are often hidden from view. Limits on nuclear cruise missiles would also be difficult to verify using NTM because they are small and because the conventional- and nuclear-armed versions are nearly indistinguishable. In addition, the INF treaty is giving new impetus to the search for cooperative restrictions on the military confrontation in central Europe, where numerical limits have been hard to agree on in part because of verification difficulties.

The United States does not have to limit itself to NTM, however. The INF treaty, as well as recent Soviet acceptance of the use of on-site inspection in a variety of arms control settings, indicates a new willingness to accept at least some cooperative and intrusive inspection measures to verify compliance with arms limitations. This chapter examines a promising cooperative way of facilitating the verification of numerical limits on weapons that has received relatively little attention: the tagging of treaty-limited items.[1] Essentially, the use of tags transforms a numerical limit into a ban on untagged items. The result is that many of the verification advantages of the zero option can be retained for a numerical limit. Moreover, tagging systems can verify a numerical limit without yielding simultaneous information on the location of all limited items, thereby reducing the intrusiveness of the monitoring required to achieve a given level of confidence that a limit is being obeyed.

Tagging works by certifying that every weapon observed is one of those permitted under a numerical limit. A tagging system would involve the manufacture of a number of tags equal to the number of weapons limited by treaty. One tag could be affixed to a crucial part of each allowed weapon. If even one untagged weapon were ever seen (by NTM, through on-site inspections, or even by nationals of the inspected party loyal to the treaty regime), then there would be *prima facie* evidence of a treaty violation. Other methods of counting a deployed force can only suggest that the allowed total is being exceeded, an indication that is unlikely to be conclusive and that might tend instead to cast doubt on all the information going into the count. Tagging produces a much stronger impetus for political action in the event of a violation, because observation of an untagged system would provide unambiguous evidence of an overall violation.

Tagging does not function as an independent verification system. Tagging would only be useful as an adjunct to NTM or as part of a fabric of cooperative verification procedures carefully tailored to a specific treaty proposal. Tagging systems have three crucial ingredients: a number of tags equal to the number of allowed weapons, a mechanism for associating a tag with a unique weapon, and a protocol for verifying the authenticity of the tags. In most applications, checking tags would be an aspect of on-site or challenge inspections, but systems are conceivable in which the authenticity of tags would be checked remotely. In some contexts, tagging systems that do not require the affixing of any physical tags may be feasible.

The chapter goes on to explore the potential value of tagging by describing the possible application of tags to five types of deployment limits. It then presents a discussion of various general problems that arise

---

[1]Tagging systems have been discussed previously by Garwin and by Fetter in, respectively, "Tagging Systems for Arms Control Verification," Report No. AAC-TR-10401/80, Analytical Assessments Corporation (Marina del Rey, California, February 1980 [Sponsored by the Office of Technology Assessment]) and "The Use of Tags in Monitoring Limits on Mobile Missiles," UCID-21034, Treaty Verification Program, Lawrence Livermore Laboratory (Lawrence Livermore National Laboratory, March 1987).

in tagging, together with possible solutions, followed by a discussion of how the complex additional burden of designing and negotiating a useful tagging regime might be borne in arms control negotiations. Finally, the question of when the benefits of verification by tags may be likely to outweigh the disadvantages is addressed.

Examples of Tagging Systems

The following examples are intended to show the weaknesses as well as the strengths of the tagging concept.

Soldiers. One possible arms control limitation is on the number of soldiers allowed in certain areas. Limits on the number of troops in central Europe have been under discussion since the mid-1950s, as have schemes for monitoring such an agreement. Usually a continual presence of inspectors or remote monitoring equipment at checkpoints supplemented by occasional forays by human inspectors has been thought to be required. It is unclear, though, how the observation of an unusually large number of troops in a particular region would be anything more than an occasion for suspicions that could not be easily resolved. Conversely, the intrusiveness required to monitor agreed force dispositions in this manner might yield evidence of local or overall force weaknesses that in a crisis could make the military balance less, rather than more, stable. With a tagging system in operation, the discovery of a single soldier without proper identification (that is, without a tag) would be conclusive evidence of a violation, yet no information need be collected about either the overall number of troops in the region or their disposition.

A tagging system for troops might work in the following manner. Suppose that limitations were imposed on the total number of active military personnel in each of several zones. At random or fixed intervals (say, every six months) the monitoring party would supply enough ID cards (tags) so that the monitored party could issue one to each soldier in the zone. The ID cards would have a section where a thumb print could be registered within two or three days of the issuance of the card. (Chemicals in the card could ensure that after this active period the thumb print would either no longer register or that the card itself would indicate that a longer delay had occurred.) Every soldier in a controlled zone would be required to carry the appropriate ID card with his or her own thumb print. Transfers of soldiers could be accommodated by the exchange of used ID cards for new ones.

With such a system in place, if an inspector ever found a soldier without a valid ID card, there would be a clear violation that could be investigated directly. In most other verification schemes, the total number of soldiers in the zone would be inferred from the number and types of units observed to be deployed there, or from some other set of imprecise measures that would not identify any specific individual as constituting a breach of the limit, even if they gave some general indication of a violation.

In practice, this tagging scheme would have to be elaborated in great detail. Most obviously, there is the technical design of an ID card that could be personalized by thumb print within the required time period and not provided only to those troops likely to come into contact with inspectors. Because the cards would be provided by the tagging party, and because inspectors could randomly recover a small fraction of the ID cards and return them to the laboratory for detailed analysis, occasional changes in the details of card technology could be used to ensure over time that no continuing copying or misuse of the tags occurred.

The example suggests several other aspects of any tagging system. As implied by the mention of inspectors, tagging only works if there is some chance of "observing" the controlled items and the presence or absence of associated tags. In the case of ground troops, it is assumed that inspectors would be given fairly free access to transit routes, if not to all military bases. The personalized quality of the ID card would ensure that no single tag could be used to provide safe transit for a succession of soldiers, who would then disappear into uninspectable bases or other safe havens.

Finally, the example suggests that although tags can help ensure that a precisely defined limit was not exceeded, there are many potential problems of verification and arms control more generally for which tags would provide no help at all. If an inspector, for example, came upon an individual in uniform with an automatic weapon but no tag, it might be explained that the person was a police officer or some other quasi-military officer (e.g., a customs agent) and not a soldier at all. In general, soldiers traveling out of uniform and separately from their weapons on civilian transport may not be identifiable as soldiers. Tagging cannot remedy imprecise definitions of what is controlled by an agreement. Only if the parties can agree on a clear definition of who is a "soldier" can numbers of soldiers be controlled.

If a precise definition could be agreed upon, however, in periods of international tension (or up to twice a year at the option of either side), soldiers as defined by the treaty might be required not only to carry tags but also to have their foreheads marked with temporarily indelible ink. This procedure would help ensure that inspectors could identify soldiers and that others not so identified could not contribute substantially to the prosecution of an attack. Over any substantial time period, soldiers could not operate effectively if they could not train, were not formed into units, and if they lacked weapons and communications. Limits on manpower, verified by tagging, accompanied by other sorts of constraints to prevent circumvention of the numerical limitation could effectively constrain military potential. If these limits were applied in many local sectors, one could build confidence that concentration for a local attack was not under way.

Tanks and Other Conventional Weapons. Because they are an essential element in offensive military potential, tanks are an obvious target for negotiated arms limitation. Compared to limits on people, limits on hardware are in some ways easier and in some ways harder to verify by tagging. Unlike people, certain major classes of military hardware have no civilian use and so cannot merely blend into the civilian landscape. Such specialized hardware includes tanks, artillery, fighter-bombers, most bridging equipment, and most munitions, but not jeeps, trucks, buses, and transport aircraft. Observation of a tank leaves little question that it is a controlled item. For the same reason of singularity, though, close-up scrutiny of a soldier would reveal fewer military secrets than close-up scrutiny of an advanced weapon.

Several other differences make hardware harder to control through tags. First, there is less reason for any particular piece of hardware to emerge from hiding or to be involved in exercises and training. An opponent determined to violate an agreement could maintain a stock of tagged equipment to be used in peacetime operations *and* an untagged stockpile that would be kept out of view until shortly before the outbreak of hostilities. This problem is conceptually similar to the possibility of unknown stockpiles in an absolute ban on a class of weapons. A possibly decisive difference, though, is that in the case of a numerical limitation, troops would have the opportunity to train with the legal weapons of the same type.

Second, each hardware item has less of an essential identity than a person. With a thumbprint tag, one can be sure that the monitored party is not using a single tag and "transplantable thumbs" to cover the transit of multiple people across inspected areas. With hardware, some care would have to be taken to design ways to take the equivalent of a fingerprint for each controlled weapon or to attach absolutely nonremovable tags to crucial pieces of the item. The complexity of the problem is indicated by the fact that a nonremovable tag on the fender of a tank would be of little help because the same fender could be unbolted and used *seriatim* to transfer large numbers of tanks to unknown storage warehouses. If the turret of a tank represented a large part of the value of the tank and the turret could not be easily removed or concealed, however, then a nontransferable tag on the turret would suffice, because a limit on tank turrets would be equivalent to a limit on tanks.

As a technical matter, a nontransferable, noncopyable tag for a tank turret is not hard to devise. One would not require absolute confidence that each tag had not been tampered with, and so one could accept the occasional reliability problems that attend any electronic components and batteries. The tagging system could use a capacitance, contact, or ultra-sound sensor, or two electrically communicating devices on opposite sides of the turret, to ensure that once emplaced it could not be removed without providing a tell-tale record. More simply, a limited amount of special epoxy glue made with unstable components and identifying trace elements or isotopes might be provided. Or tags might be emplaced with ordinary glue and an ultrasound fingerprint of the resulting assembly recorded. In the case of an electronic tag, copying could be prevented by crypto-graphic keys stored in a shielded microchip. Tags that would be recovered occasionally could be made further secure against copying through serial numbers and the recording of random aspects of their physical micro-structure, through the use of minute amounts of complex artificial chemicals, or through the use of altered isotopic composition in specific small parts of the tag. A nation attempting to counterfeit such tags could never be sure that the copy duplicated all the identifying characteristics of the tag.

If several classes of tags were provided, or if the tags had serial numbers, then tagging could be used to control the number of tanks in each of several zones of interest, as well as the total in the overall region. The tags for each zone might be different colors and shapes, so that close inspection would not be required to ascertain that a tank was in an allowed area; only a small number of random close-in inspections would be required to verify that the tags were authentic. Such a scheme would be complicated, though, if tanks were rotated among zones and new tags were thus required to be installed.

If details about tank dispositions were not considered sensitive, the monitored party might simply be responsible for turning over to the monitoring party a roster of which tag serial numbers were in each zone prior to the beginning of each inspection period. Even if the dispositions were sensitive, the roster idea could be adapted, using cryptographic techniques, so that the monitored party could keep the overall roster secret while still providing assurance that any particular observed system was within a sub-limit for a particular zone. Cryptographic or electronic means could be used to produce the equivalent of a system where the roster is deposited with a neutral and confidential judge who responds "yes" or "no" to queries of the form, "Is tank number 1197 allowed in zone 11?"

Rail- or Land-Mobile ICBMs. ICBMs are larger and more valuable than tanks, and they probably require more frequent servicing. These may not be the decisive differences where verification is concerned. Nations may be more anxious to keep secret the technical details of their construction, and aero-dynamic requirements are such that tags probably could not be permanently attached to missiles. Tagging the canisters in which the missiles are transported and stored is an obvious alternative, but this would require that the tag provide information sufficient to ensure that a tagged canister could not be used to transfer illegal missiles to a covert deployment area. The tagging scheme would have to ensure that canisters were not returned to the factory or repair depots empty or containing a decoy. This could be accomplished by an agreement that missiles would not be removed from canisters except at designated repair depots equipped with appropriate portal monitoring and inspection systems. A seal or an acoustic sensor that monitored standing waves inside the canisters could be used to ensure that the canisters were not opened except at designated facilities.

Tags on missiles (or missile canisters) could be checked in a variety of ways. In the case of a rail-mobile missile system, it may be the case that such trains would have a distinctive and undisguiseable signature; the length or weight distribution of the trains, for example, could be measured using an unmanned sensor. After identifying such a signature, a tag reader, using a short-range radio or infrared beam, could try to interrogate a tag. Such automatic systems might be installed at choke points in the rail network.

For land-mobile missile systems, the verification protocol might allow occasional free access by inspectors within a random fraction of the specified deployment area. Tags would be checked on any missiles that were found in the area. Such inspections would be relatively effective. Consider, for example, a case in which 100 untagged missiles were illegally placed in a deployment area. If each inspection examined only one percent of the deployment area, and if there were only a 20 percent chance of locating each missile present in the inspection zone, then only four inspections per year would assure an 80 percent chance of discovering the violation within two years, no matter how many missiles were allowed under the agreement.

Alternatively (and perhaps especially outside of agreed deployment zones) the parties could rely on NTM to keep track of controlled systems and their tags. A tag incorporating a navigation system (either radio or inertial) might record its own movements, eliminating the need for real-time access for direct inspection. If the movements of the mobile system were randomized, then knowing some of the past movements of a weapon would not compromise its future survivability. If there were questions about a controlled item observed by NTM on a highway at a certain time, it would be possible, by examining the stored movement record, to prove whether or not a tagged system had been at that location at the time in question. The parties might even be willing to give each other constant information on the location of tagged missiles outside of deployment areas, reducing the risk that the monitoring party would reveal information about its intelligence capabilities in the process of requesting challenge inspections. The characteristics of the controlled weapon and the detection system would have to be such that a tagged weapon could not serve as a decoy to provide safe passage to one or more untagged weapons kept in close proximity to a tagged system. Tagging cannot eliminate the possibility of unknown deployments of limited weapons. If inspectors and tag readers were placed at the portals of all known production, test, and repair facilities, however, one could be confident that no untagged weapons would be serviced there, thus forcing a cheater to establish a completely parallel covert maintenance and testing system. The risk of clandestine deployments could be further minimized by limiting the number of people trained to operate the controlled system.

Cruise Missiles. Weapons that are relatively small and easily moved are of course less likely to be observed by national technical means, either in normal operation or when deliberately concealed. Limitations on cruise missiles could be verified with tagging schemes similar to those used for mobile ICBMs, but cooperative measures and more stringent inspections would be essential to increase the likelihood of detecting untagged systems. The key locations to monitor a limit on cruise missiles would probably be repair and maintenance facilities.

The dual nuclear and conventional capabilities of cruise missiles also pose problems. If there is concern about ostensibly conventionally-armed cruise missiles actually being deployed with nuclear warheads, tags might be modified to include plastic scintillation material and appropriate instrumentation, which would provide evidence of any close-by nuclear material. Of course, even if a tagging scheme were capable of detecting the peacetime deployment of nuclear warheads on missiles declared to be conventionally-armed, it could not prevent replacing conventional warheads with nuclear warheads in a crisis or at the outset of war. One might solve the quick-conversion problem, however, by agreeing to specific technical restrictions on cruise missile design.

Submarine Deployments. Coastal keep-out zones in which missile-launching submarines would not be allowed have been suggested as a way to guarantee increased warning time for command authorities and for alert strategic forces, thereby reducing a possible incentive to strike first during a crisis. Alternatively, safe zones from which attack submarines and other anti-submarine warfare forces would be barred have been suggested to improve the survivability of the missile-launching submarine forces. Although compliance with such agreements in peacetime obviously could not guarantee their continued observance in wartime, it is worth noting that a tagging scheme could allow nations to verify compliance with such agreements even during crises.

Each side would be given a limited number of challenge opportunities each year during which it could ask a specific submarine to demonstrate that it was outside the prohibited areas. Because submarines cannot travel fast and remain undetected, it would be sufficient for the submarine to surface and make its location known within two days, so long as it surfaced more than perhaps 1,000 miles from the keep-out zone. A transmitting tag, which would be carried permanently on-board each submarine but which would be turned on only during such challenges and only after the submarine surfaced, would serve to identify the submarine as the one whose position had been requested. The time delay would reduce the information about patrol patterns that might otherwise be gleaned from these inspections.

General Characteristics of Tags

The idea underlying any tagging scheme is that if all allowed items are tagged, then the detection of a single untagged item would constitute direct evidence of a treaty violation. In principle, tagging makes monitoring numerical limitations as easy as monitoring total prohibitions, because the monitoring party need only verify the ban on untagged items. On the negative side, however, any tagging system itself would introduce a degree of complexity to arms control verification, and the continued existence of allowed production, testing, and operational capabilities may prompt worries about potential evasion of the monitoring system or a sudden break-out from limitations. Still, tags offer considerable potential for improved verification regimes.

Any tagging system should have the following general characteristics:

1. It must be impossible to copy the tag without detection, for otherwise the monitored party could simply produce counterfeit tags to cover weapons deployed in excess of the limit. To make it more difficult for the monitored party to learn how to copy tags, the tags could be replaced at intervals with ones using different anticounterfeiting techniques. As a test before a tagging system was agreed upon, the monitoring party could offer a prize to any citizen who succeeded in defeating the anticounterfeiting scheme.

Tags might be made non-copyable in three generic ways: use of coded electronic signals, use of some natural property of a material for identification, and use of artificial properties that need not be fully disclosed to the tagged party.

Electronic tags have many advantages: the technology is well understood, the cost is likely to be low, the identity of a particular item need not be divulged, and the authenticity of a tag could be determined without direct access to the weapon on which it is emplaced.

If parties to an agreement do not object to tags that would identify individual weapons, then the simplest electronic tag would work as the equivalent of a "one-time pad." Electronic access to the tag's memory would only be allowed following the input of a special code unique to the particular tag and to the number of times it had been read previously. Each tag would report its serial number, the number of times it had been inspected, and a unique secret number for that serial number and index. The secret numbers would be compared with a master list to authenticate the reading. Each secret number would be erased after it was read, and the series of secret numbers would be different for each tag. The monitored party could be informed of, or learn by its own devices, all the information that was transmitted to and from the tag during this process and yet could not use this information to counterfeit tags.

If parties were unwilling to allow the identification of individual tags, then more complicated cryptographic schemes would be required. If the tags cannot indicate their identity, then their input and output must be identical. If tags themselves are identical, then the problem of preventing illegal duplication would be a difficult one. In either case the tag would have to be protected by various physical means against nonelectronic means of discovering the series of secret numbers; for example, the chip could be shielded and provided with a membrane that would trigger a self-destruct mechanism if violated. In the case of unique tags, each tag need only protect its codes against tampering that does not leave any indication of misuse. But in the case of identical tags, one must prevent even destructive means of discovering the secret codes, because a few tags could be sacrificed in a counterfeiting effort. This may be possible; the tag need only destroy its information in response to intrusive examination. Another possible method of allowing the parties to retain the anonymity of particular weapons would be to allow the use of unique tags but to interpose a piece of sealed equipment or a neutral party between the signal from the unique tag and the tagging party. This intermediary would certify that the tag reading was valid but not reveal the detailed basis of this certification.

Tag reading could be done by a direct connection to a local or remote console, or the tag could be queried by a transmitter. In the case of unique tags, especially those in which the tag or tag reader records the geographic position of the tag, the most sensible approach is to allow the monitored party to provide the communications circuit from the monitoring party to the tag.

Tags could also be based on patterns in a certain material or substance; for example, one could take a three-dimensional image of a certain portion of a fiberglass missile canister using a stereoscopic camera, an acoustic or electron microscope, or acoustic holography. Alternatively, identifying material could be affixed to the weapon being controlled, as with glitter blown into a layer of epoxy on the weapon, or the use of a fiber-optic seal. If the tag were an intrinsic characteristic of the weapon, which would make duplication, spoofing, or swapping difficult, the tagging would have to be done on-site by the monitoring party. The tag reading would be done with the same type of instrument used for the initial imaging, which would almost certainly require an on-site inspection. The pattern itself could be public knowledge, because the principle of the tag in this case is the irreproducibility of complex three-dimensional patterns. One would only need to be sure that the pattern came from a particular tag. Although all tags based on patterns would be inherently distinguishable, a tag reader could be devised to convert the identification information into a "yes" or "no" answer.

As noted already, there is great scope for using very subtle features of tags to prevent their duplication if some fraction of the provided tags could be recovered and tested in a laboratory. Such subtle features could include altered isotopic composition of particular parts, the deposit of a monoclonal antigen within a fiber, or seemingly random imperfections in a printing or manufacturing process.

2. It must be impossible to spoof the tagging system or to fool it into thinking that a valid tag exists where there actually is none. For example, it must be impossible to reroute signals between the tag reader and a counterfeit tag so that the tag reader would actually receive a return signal from a valid tag at another location. Although preventing or detecting such signal displacements would be straightforward if an inspector had direct access to the tag, special precautions would have to be taken if tag reading was accomplished remotely. The general solution is to include coded location and time information in the response elicited from the tag.

3. It must not be possible to move the tag from one weapon to another without the knowledge of the monitoring party. If tag swapping were possible, then valid tags could simply be moved to the weapons being inspected at a particular time and place, or at least to those systems more susceptible to inspection by the other side. If tags were glued onto the tagged weapons, it could be arranged for part of the tag to change color or melt if exposed to the solvent required for the glue employed. An analog is in use in the U.S. domestic economy: to discourage the illegal parts business, automobiles now are made with serial number tags glued to their major sheet metal parts, and owners are warned not to attempt to remove these labels.

4. The tagging system must not aid the monitoring party in locating weapons in real time, because this could render tagged weapons more vulnerable to preemptive attack. Such position information might even allow terminally-guided munitions to home on the tags during an attack. For example, a radio beacon attached to tanks or mobile missiles would certainly allow them to be counted by satellite receivers, but it could also allow attacking warheads to home on those targets.

5. More generally, the tag should reveal only information that is required for the purposes of verification. In other words, tags should not be agents of espionage that collect sensitive data about the limited weapon

or its deployment patterns. Parties might be unwilling, for example, to emplace tags that could reveal low rates of readiness previously unknown to the other side. Concerns about espionage could be alleviated if the physical details of the tags and the tagging system's operation were exhaustively disclosed to the monitored party, but this would restrict the use of sensitive technologies and may make the tags easier to copy or spoof. On the other hand, the use of open-tag technology would make it easier to publicize evidence of treaty violations, because no sensitive sources or methods could be compromised. The monitored party could be assured that the tags are what the monitoring party says they are by providing twice the number of tags, half of which could be selected at random, disassembled, and returned. It would be impossible to verify that there were no secret aspects of the tag (as noted above, some subtle secret aspects would be useful to prevent counterfeiting), but it should be easy to verify the absence of homing devices, chemical explosives, cameras, or other intrusive devices.

In some contexts it may be desirable that tags not uniquely identify particular weapons. The monitored party may be concerned, for example, that valuable information could be gained if the monitoring nation were able to trace the deployment history of individual weapons. Although the easiest way to make tags irreproducible is to give each tag a unique serial number, other approaches could also be used to prevent counterfeiting.

6. The tagging systems must be extremely reliable and have a very low false-alarm rate. False alarms not only undermine the mutual trust of parties which a treaty otherwise might engender, but, in sufficient number, they could create a background against which cheating would become easier. Designers of tagging systems should give some attention to reducing the possibility that the monitored party could deliberately act to increase the false alarm rate as a prelude to an episode in which illegal weapons would appear in transit or in repair and then be concealed.

7. The physical size and power requirements of the tag should be such that the normal functioning of the tagged weapons would not be impaired in any way. Once again, the use of open tag technologies combined with the random inspection of tags should reassure the monitored party that the tag could not somehow harm the weapon.

8. The tag must be reliable in the full range of environments that the weapon might experience during storage, testing, training, repair, and deployment. These factors may include extremes in temperature, vibration, humidity, radiation, etc., and some degree of deliberate abuse or tampering.

9. The tagging system must not be excessively costly. An acceptable hardware cost might be as much as a few percent of the cost of the limited weapons, especially if operating costs can be kept relatively low. Because a small percentage of the cost of a major weapon system could amount to tens or hundreds of millions of dollars, it seems likely that effective systems can be designed within this constraint.

Verification Systems Using Tags

All verification procedures seek to raise the political risk, increase the technical difficulty, and elevate the economic cost of cheating. No system can eliminate all possibility of cheating, but cheating can at least be made risky, difficult, and expensive. For example, tags could not discover hidden stockpiles of undeclared weapons, but they could make it impossible to mix those weapons with weapons being counted against negotiated limits. Depending on the facilities that would be open to inspection, this would force the monitored party to develop a completely parallel but covert system of production, assembly, storage, testing, training, repair, and deployment of its secret stockpile. Not only would the economic cost of such covert stockpiles be much higher than that of allowed weapons, but the risk of being caught—simply by an accident that exposed an undeclared weapon to the light of day—could well outweigh any military advantage that might otherwise have been gained from the undeclared inventory. If, for example, the testing of covertly-produced missiles could be prevented (such testing is easily monitored by NTM), then covert missiles would become much less valuable to a potential cheater. Large clandestine facilities would probably be required to maintain, test, and store hidden stocks, because it is generally agreed that the size of any undeclared inventory must be a sizable fraction of the allowed inventory before it would be significant militarily.

The following section examines more closely how tagging systems might operate. The discussion is organized according to how the tags would be checked—by on-site inspectors, through remote telemetry, or at natural choke points or artificial portals in the monitored country.

Tags as an Aid to On-Site Inspection

Perhaps the most straightforward way to use tags would be in conjunction with on-site inspection. The use of tags would provide a clear way in which information gained at individual on-site inspections could contribute to an overall judgment concerning compliance with a treaty. Without tags, on-site inspections cannot produce much direct information about the total number of weapons deployed unless all sites are inspected simultaneously. Simultaneous inspections not only would be extremely intrusive, but, for many types of weapons, simultaneous revelation of the location of all weapons would also raise the specter of a preemptive strike. In addition, simultaneous inspection would require great numbers of inspectors and host-country guides.

Under more plausible inspection schemes, but without tagging, one might learn that thirty missiles were at site A in January, forty at site B in June, and fifty at site C in December. There would be no inherent way to conclude whether or not the total number of missiles at all sites at any one time exceeded the permissible limit. A periodically declared roster of how many missiles were at each site would reduce this problem, but such a roster would not give confidence by itself about the completeness of the count at a given site. By contrast, if all allowed missiles were tagged, one could tell if every missile found at whatever facility was part of the allowed inventory. A single missile found anywhere without a valid tag would be *prima facie* evidence of a treaty violation. The first step in instituting a tagging system is to affix tags to the controlled weapons. This could be done during an initial round of on-site inspections, or, if the anti-swapping measures were sufficiently foolproof, one could simplify the process greatly by passing out the allowed number of tags to the monitored party, whose own personnel would affix them. Ideally, tags should be designed so that they could only be affixed within a short time after the monitoring period began, or at least so that when inspected they would give some evidence of how long they had been attached. There should be strong incentives for the monitored party to affix the tags promptly and properly to avoid a reservoir of tags that could be affixed to weapons that happen to be selected for inspection. In significant degree the necessary incentive is inherent in the tagging scheme, because maintaining a reserve of tags would increase the number of untagged weapons and thus the chance that an untagged weapon would be discovered. Note that tags need not be irremovable—they must only indicate in some obvious way that they had been removed.

Careful thought should be given to the particular component (or components) to which the tags would be attached. The component should be an essential part of the weapon system, and it should be difficult to swap this component between systems on short notice. As noted above, the main turret would be a good place to tag a tank, and the barrel might serve for an artillery piece. Many missiles are normally stored in canisters, and most mobile missiles are launched directly from the canister. Although one would naturally prefer to tag the missile itself, this would entail reading the tag through the canister or opening the canister for inspection, either of which might present difficulties. If the canister were tagged but not sealed, allowed missiles could be swapped with undeclared and thus prohibited missiles, thereby providing undeclared missiles access to declared facilities. Sealing the canister would almost certainly require an on-site human presence, however, and would probably complicate missile maintenance. One could possibly develop a fiber-optic mesh that would surround a canister, while still allowing access to small missile components inside for adjustments and repairs, but not allowing separation of the canister and missile. The mesh would be made of a continuous single fiber that could not be cut without interrupting a light beam flowing through it, giving a signal that would be recorded by the tag electronics.[2]

---

[2]The fiber-optic mesh was suggested to the authors by Richard L. Garwin.

During an on-site inspection, inspectors would locate limited weapons and attempt to verify the authenticity of their tags and to verify that the tags had never been removed. (If tag reading was difficult, as would be likely for pattern-based tags, a random sample of the tags could be checked.) Electronic tags could be equipped with low-power infrared transponders (much like a television's remote control), thereby allowing the tags to be queried from a few tens of meters away. Such tags are already in use in commercial assembly lines for inventory control. This would reduce the intrusiveness of on-site inspections and yet not provide a homing capability that would make the tagged weapon vulnerable to attack. An extension of this idea would be to let robots inspect the tags or to fly a pilotless airplane over the site to query tags.

Procedures would have to be worked out for the return of a tag when a controlled weapon was destroyed or otherwise removed from the inventory. To prevent testing of undeclared missiles produced at covert facilities, one would need to verify that the missile being tested had been tagged and thus had been taken from the allowed stockpile. Tags would alleviate the need for detailed monitoring of other methods of destruction.

If the verification regime permitted inspections on short notice at the option of the monitoring party, they could be timed to take maximum advantage of national intelligence capabilities. The movement of controlled weapons into or out of the facility to be inspected could be monitored closely by NTM or by special cooperative measures just prior to the event. Ideally, the facility would be closed or put into a stand-down condition until the completion of the inspection. If inspections could be conducted on short notice, the movement of any illegal weapons out of declared facilities might be detected. Even if an untagged missile were never actually found during an on-site inspection, tagging could force a cheater into more obviously suspicious behavior. Moreover, because untagged weapons would have to receive special handling at all times, tags might greatly increase the number of people who knew of a treaty violation on the part of their country, increasing the likelihood that the violation would become widely known.

The use of on-site checking of tags in providing evidence of cheating—indeed, the use of any type of on-site inspection for this purpose—should not be oversold, because access to the evidence would always be in the control of the monitored party. Although it is true that the detection of a single untagged missile would be evidence of a violation, the monitored party would be unlikely to allow an on-site inspection when such a possibility existed. It would always be advantageous from the cheater's perspective to make up excuses for delaying or denying an on-site inspection rather than risk discovery of a "smoking gun." This may lead to a paradox of sorts, because if a tagging system were implemented, the lack of a tag could become, in the eyes of the world community, the *only* acceptable evidence of a violation.

Thus, even though tags could provide unambiguous evidence of a violation with just a single observation, it is unlikely that this would ever happen during an on-site inspection. The monitoring party probably would have to act on more ambiguous evidence, such as a refusal or delay of on-site inspections, surreptitious movement of missiles out of declared facilities, tag tampering, or other suspicious behavior. Tagging would have played a role, however, in elucidating this suspicious behavior. Moreover, because of tagging's relative efficiency in detecting violations, tags should reduce the likelihood that a country would decide to cheat in the first place (which is presumably the main purpose of verification).

Tags read on-site could be an excellent way to help build confidence between parties who are in compliance with an agreement. Because tags make inspections more effective, they would have the virtue of minimizing the number of inspections required for a given level of confidence. Tags also could reduce the chance that false claims of treaty violation would be used for political reasons.

Monitoring Tags Remotely

In general, verification regimes are likely to be easier to negotiate if requirements for on-site inspections, especially those involving trained foreign personnel at sensitive military locations, are minimized. If tags could be read remotely, routine on-site inspections would not be needed to verify limits on even small, concealable weapons. Three basic schemes using remote reading come to mind: the tag could transmit a continuous or intermittent signal, the tag could be provided with a two-way communication link, or the tag could record position information for later interrogation.

The most obvious remote sensing method is for every tag to transmit a coded set of high-frequency radio pulses. The location of the tag could then be determined by satellite receivers using time-of-arrival measurements. If other arrangements are made for tracking tags outside of deployment areas, the power requirements for the tag beacons could be kept low by installing a set of time-of-arrival receivers and a satellite earth station in each deployment area. The obvious drawback of this scheme is that one might be able to home on the beacons during an attack. The monitored party might be given the ability to switch off the transmitters in time of crisis to ease this problem, but this would not eliminate the possibility of a surprise attack. In addition, such a system might aggravate a crisis, because switching off the beacons could be taken to indicate that the monitored party was preparing for war. Even worse, there could be pressures to launch an attack while the beacons were still on, or shortly after they were switched off, when the approximate location of the tagged weapons would still be known. The very necessity of making such decisions would distract leaders from dealing with more substantial issues.

A better plan would be to have the beacons emit signals randomly and infrequently in time, so one would never know the location of a large fraction of the tagged weapons at any one time. An inventory of weapons, for example, could be equipped with beacons that emitted a signal once every ten days. If the weapons were moved once per day, then the monitoring party would only know the location of 10 percent of the inventory at any one time.

In another remote-monitoring scheme, each tag would contain a receiver that recorded position information given by a navigation system. This system has the advantage that the quality of the location information could be controlled. If, for example, the resolution of the navigation system is too great, then the system's output could be filtered to report only the number of a map square in which the tag could be found. After a period of time, the degraded information stored in the tags could be transmitted to the monitoring party. This transmission could be encrypted and security codes added to ensure the authenticity of the data. If the time delay were short (a few days), this idea would be similar operationally to the beacon scheme. Alternatively, the tags could be collected and sent back to the monitoring party and new tags issued. The tags themselves would then constitute a time-lagged data base of the position of every allowed missile. Tags of this type could be used to enforce regional limitations on weapons, such as the number of tanks near the central front in Europe.

Of course, neither of these tagging systems could detect undeclared weapons. The presence or absence of undeclared weapons would be verified by comparing the location information supplied by the tags to NTM data. For example, a satellite photograph that showed a controlled weapon at a location that was not recorded by any of the tags at that time would be evidence of a treaty violation. The advantage of this method is that it would provide reliable data on the variable of interest: the number of allowed weapons. The cost would probably be low, and data-handling requirements for this system would not be excessive.

These systems do not resolve all problems, however, and they create some of their own. First, they rely on NTM to detect violations. Because a cheater would be careful not to expose undeclared weapons to reconnaissance satellites, the probability of observing a violation would be small. One would probably have to depend on accidents (e.g., the crash of a train carrying covert missiles) to expose or deter cheating. Second, the system would place high demands on technology. It may not be possible to build the type of tag described here—the receiver or beacon may simply be too large or require too much power. One may have to develop a new receiver and perhaps a new navigation system, which would increase costs greatly. It also may not be possible to develop tags that are sufficiently reliable. If a photograph shows the location of a controlled weapon but the tag records the position information inaccurately, then a false indication of a treaty violation would occur. Error-checking and validation protocols may be able to reduce the false-alarm rate to negligible levels, but this would have to be demonstrated. Third, the monitored party could not program the movements of tagged weapons on fixed schedules because the monitoring party would quickly learn these patterns and be able to predict the location of, and therefore target, the weapons in the future. This is a minimal concern. Prudent planning already requires that deployment patterns be random, with or without tags, because mobile missiles depend for their survivability on the opponent not being able to predict where they will be at any given moment.

Monitoring Tags at Choke Points

Tags also could be used effectively at natural or artificial "choke points," or places through which all or most of the declared weapons must pass at least occasionally. As an example of a natural choke point, consider a limit on rail-mobile missiles such as the Soviet SS-24 system. It is likely that a number of missiles would be deployed on the same track, or at least that several missiles would have to pass a certain point on the track to exchange positions (the position of choke points would depend on the topology of the rail network). If choke points could be identified, tag readers could be installed at these points, along with sensors to detect untagged weapons. Imagine, for example, that a missile were approaching a choke point equipped with sensors. If the missile had a valid tag, the sensors could read the tag (perhaps using the infrared transponder mentioned earlier). If an undeclared missile tried to pass through the choke point, however, other sensors, such as scales or x-ray machines, would determine that the object could be a missile. The monitored party would then be required under the verification regime to allow more intrusive inspection to prove that the object was not a limited weapon (e.g., video cameras could be used to look inside the railroad car). A refusal to allow such an inspection would cast serious doubt on treaty compliance, although it would not constitute direct evidence of a violation.

Another example of a natural choke point can be found in the deployment of nuclear-armed cruise missiles on submarines. Because the portals for bringing cruise missiles on board submarines are likely to be limited, one could deploy sensors that would detect the presence of fissile material at each portal. Every time the sensors detected fissile material, they would also expect to read a valid tag. This scheme is clearly not foolproof, but it might be better than allowing the number of submarine-based cruise missiles to remain unrestricted. This scheme is unlikely to work with surface ships, because there are too many ways to bring missiles (and heavily shielded warheads) on board. (Given the possibility of underway replenishment of submarines, the scheme may be unworkable for them as well.)

If a natural choke point could not be found, one could be created by surrounding declared facilities with monitored fences that force the movement of mobile missiles or critical components through a gate where they could be observed and counted. The declared facilities could be any combination of production, assembly, storage, testing, training, repair, and deployment areas. The fence, or perimeter, would be a two-dimensional barrier around the monitored party's facilities that could not be violated without detection. A wide variety of fence sensors could be used, including seismic detectors, microwave intrusion detectors, acoustic sensors, video and infrared cameras, metal detectors, short-range radars, or pressure sensors. Possible monitoring devices at the gate, or portal, might be video or infrared cameras, weighing scales, x-ray, gamma-ray, neutron, or ultra-sound imaging devices, metal detectors, and human inspectors. The perimeter/portal data could be transmitted to the monitoring party in a secure mode or interpreted by human inspectors stationed at the site.

Consider the case of a perimeter/portal system at an assembly plant. If a limited weapon had not yet been produced, this would be the ideal point to verify limits on the weapon so long as it could not easily be assembled without detection at other unidentified or undeclared facilities. When a finished weapon was ready to leave the assembly plant, the monitored party could simply declare the weapon and the count of deployed systems would be increased. If the monitored party did not declare the weapon, monitoring devices at the portal would determine that the object *could* be a limited weapon. Unless further inspection were permitted to determine that the object was not a limited weapon, the monitored party would be in violation of the treaty when the weapon left the facility.

This system would have the advantage that declared weapons would not be inspected by intrusive devices at the portal. But to retain this advantage, declared weapons would have to be tagged before leaving the facility so that they could be returned for maintenance. Without tags, the monitoring party would have to inspect any returned weapons to ensure that the monitored party was not returning bogus weapons and replacing them with real weapons. Tags would also prevent covertly-produced weapons from having access to declared production and assembly plants.

It is much more likely, however, that a substantial number of weapons would already have been deployed before limits could be placed upon them, in which case a method to establish the initial inventory would be needed. On-site inspections at declared facilities could help establish the initial inventory, or perimeter/portal systems could be constructed at these facilities. In the latter case, all existing allowed weapons would be tagged. The tag would be queried at the portals of testing, storage, training, repair, or deployment facilities, and only weapons with valid tags could enter the facility. As before, sensors at the portal would detect any undeclared object that could be a limited weapon; such objects would be subject to further inspection or be denied passage within the terms of the treaty. If desired, weapons could be monitored even while in transit between declared facilities by installing tag readers along commonly-traveled routes or by attaching a tag containing a navigation receiver or inertial-guidance package to the weapon.

In the absence of tagging, deployment areas present a serious problem for perimeter/portal systems for most weapon systems: the perimeter would have to be very large and therefore expensive to instrument. For example, if there were ten deployment areas of 100 mobile missiles each, and the missiles and launch vehicles were hardened to an over-pressure of five pounds per square inch, the total perimeter length would be at least 2,000 kilometers. Instrumented fencing may cost a million dollars per kilometer to build and install, with the entire perimeter system requiring the expenditure of billions of dollars. In addition, many of the sensors considered for the fence, especially seismic detectors, radars, and video and infrared cameras, are unlikely to be allowed unless the monitored party could be absolutely sure that the information collected could not be used for targeting. In such deployment areas, remotely-read tagging systems (or those requiring occasional access by tag-checkers) could make a huge difference in the cost and reliability of counting controlled weapons.

There are several disadvantages to perimeter/portal systems, especially when they are applied to different types of declared facilities. First, both sides may be reluctant to allow the other to construct a perimeter composed of a wide variety of sensors around some of their most sensitive military areas and to allow intrusive inspections of any entering or exiting objects that the monitoring party claimed could be a limited weapon. The potential for gathering intelligence information that was not required for verification purposes would be obvious. Second, the perimeter/portal systems would be expensive—even more so if supplemented by a human presence. Third, such a system would necessarily be complex, requiring perhaps hundreds of agreed rules governing the interpretation of data. Finally, perimeter/portal systems probably would disturb the normal functioning of declared facilities. Tagging does, however, provide a natural complement to perimeter/portal systems, allowing reduced intrusiveness within the controlled facilities.

Tagging and the Negotiation Process

Any tagging system must be carefully designed to fit both the characteristics of the weapons being controlled and the degree to which the parties are willing to divulge certain types of information (such as past position information). Because some aspects of tagging are likely to be technically complex, tagging could introduce a further element of difficulty into arms control negotiations. This technique, which is intended to increase confidence in treaty compliance, could have the opposite effect if the hardware and protocols were not devised with great care. The very act of introducing the possibility of tagging into a negotiation could delay agreement. Substantial research and development on tagging, together with focused technical discussions among the potential parties to an agreement, may be necessary in advance of any attempt to include tagging in the negotiations aimed at a specific limitation.

General Secretary Mikhail Gorbachev has suggested that a special Soviet-American committee of scientists could put forward their views on verification to the leadership of the United States and of the Soviet Union. Although setting up such a committee could be an important step, a high degree of confidence in a proposed tagging scheme might be attained only if the prototype hardware were developed and subjected to severe testing substantially in advance of an agreement. Such activities are clearly within the charter of the U.S. Arms Control and Disarmament Agency. A relatively small amount of money, on the order of $10 million, would be required for such a technology demonstration program.

This paper has focused on tagging systems for bilateral U.S.-USSR or NATO-Warsaw Pact agreements. Yet, it also would be possible to adapt tagging to truly multilateral agreements. Various aspects of the tagging protocol and of methods for ensuring noncopying of tags would be more complex, but preliminary investigation suggests that the difficulties would not be insuperable.

Conclusions

If negotiated limits on relatively small, easily-concealed weapons such as mobile or cruise missiles are important, the problem of verification will have to be solved before agreements can be completed. In general, there are three ways to go about this: ban the weapons altogether, accept a lower standard of verification than for large, fixed systems, or develop new monitoring techniques to provide adequate verification.

The first solution may be unacceptable when the weapons in question are considered to make a positive overall contribution to national security and international stability, as in the case of mobile missiles, or when dual-capable systems are already deployed, as in the case of cruise missiles. The second solution is also widely regarded as unacceptable. Many U.S. politicians are predisposed to believe that the Soviet Union will cheat on agreements whenever possible, and it is unlikely that an important treaty could withstand these suspicions unless a convincing case could be made that Soviet compliance would be verified and violations detected.

Tags could be part of the third solution. Although dozens of ideas for tags already exist, it is probably not wise at this point to spend too much time or money developing and testing tag hardware. Tags could be designed that meet all of the generic requirements outlined above: resistance to counterfeiting, spoofing, swapping, espionage, homing, etc. Instead, more work is needed to explore the feasibility of tagging concepts and to define the overall verification system of which tags could be a part, because the tag technology needed will depend much more on the verification regime as a whole than on any general requirements that tags must meet. Once a promising verification system is defined that requires a certain type of tag, then the development of specific tagging hardware could go forward productively.

Three generic tagging concepts have been considered in this chapter: tags read during normal on-site inspections, tags that give location information remotely, and tags read at natural or artificial choke points. Each system would require a different type of tag, ranging from microchip tags with infrared transponders to navigation receivers and fissile-material detectors. Using tags as a supplement to on-site inspection may be the simplest system to implement because it places low demands on technology. Tags make on-site inspections more efficient and effective and may also make them more acceptable by replacing humans with sensors of limited and known capacities, thereby decreasing the potential for espionage. Remote reading of tags further decreases the necessity for an on-site human presence, but places higher demands on technology and may be less effective because of the reliance on NTM to detect undeclared weapons. Using tag readers at choke points is an attractive idea, but it is often difficult to find natural choke points, and constructing artificial choke points could be intrusive and expensive. The power of the tagging concept is such that permanent choke points and perimeter/portal systems may be obviated.

Tags are a technical fix that will only aid the negotiating process to the degree that those technical difficulties with verification that tags could ameliorate are delaying the completion of treaties. Even if tags could make numerical limits on certain weapons easier to verify, there may be other barriers to agreement. To the degree that this is the case, instead of being part of the solution tags could become part of the problem—a source of endless detailed technical discussion that could be used to obfuscate more fundamental differences. An agreement incorporating tags would undoubtedly be far more detailed and more difficult to negotiate than one without tags. Although the United States and the Soviet Union have shown an ability to negotiate technically complex treaties—SALT II, the INF treaty, and the agreement limiting peaceful nuclear explosions, for example—such complications should only be introduced when an agreement would be impossible without them.

In summary, although tags are not a panacea for the problems of monitoring numerical limits on concealable weapons, they could have much to offer if part of a carefully-designed system. To be truly available for inclusion in a future treaty, tagging systems will have to be the subject of detailed previous discussion among the parties and should include parallel technical research and development on both sides.

───┤4├───

TECHNICAL MEANS OF VERIFYING CHEMICAL WEAPONS ARMS CONTROL AGREEMENTS

BY FRANKLIN E. WALKER

In the past several years there has been an alarming resurgence in the use of chemical weapons. The chemical attacks by Iraq on Iranian troops in 1985 is one example. The apparently indiscriminate use of chemical warfare (CW) agents by the Soviet Union in Afghanistan is another. It also has been alleged that Vietnam has used toxins in both Laos and Cambodia. Moreover, analyses of Soviet activities in the CW field indicate that the Soviets have stockpiled large quantities of CW agents and are prepared to conduct warfare in a chemical attack environment.

These considerations, together with the continuing spread of capabilities to produce chemical agents among Third World countries, have convinced the U.S. government that a verifiable treaty to prohibit the production and use of chemical agents and to require the destruction of existing stocks would be desirable. In fact, one of the few concrete agreements reached by President Ronald Reagan and General Secretary Mikhail Gorbachev at their first summit meeting in Geneva in November 1985 was that new efforts should be directed toward the negotiation of a CW treaty, and some substantive progress has been made since.

Negotiations for a new general CW treaty have been under way for many years in the Conference on Disarmament (CD) at Geneva, but until recently progress has been disappointingly slow. Of primary concern is the ability of the United States to verify compliance with any treaty negotiated. A major impediment to more rapid progress has been the Soviet Union's rejection of the specific on-site inspection procedures proposed by the United States. The USSR has accepted in principle the need for on-site inspections, however, and there has been some movement in the talks. Even so, the problem of reliable verification goes beyond these specific disagreements and derives in part from deficiencies in the preparations made so far by the United States for verifying a CW treaty. There appears to be a consensus among several government agencies that even if the U.S. proposals were accepted fully by the USSR, the detection and analysis capabilities that would be available to the United States for monitoring compliance with a CW treaty would be distressingly inadequate.

55

This chapter addresses the urgent problem of the United States' capability to monitor compliance with a CW treaty and explains how the nation could develop instrumentation and procedures to achieve an effective monitoring capability consistent with the Convention on the Prohibition of Chemical Weapons proposed by the United States. Specifically, it (1) describes relevant aspects of the proposed convention, (2) summarizes the recent negotiations on verification and monitoring issues, (3) describes relevant technologies and their derivative instruments and devices that could be used to help monitor provisions of the convention, and (4) assesses the risks and liabilities in using the suggested monitoring capabilities. It concludes by recommending a U.S. commitment to verification research and development. The details of a *specific* plan for developing the required monitoring instrumentation and procedures and for their testing, evaluation, and coordination is left to appropriate executive and congressional agencies.

The U.S. CW Convention Proposal

On April 18, 1984, the United States presented to the Conference on Disarmament its proposed Convention on the Prohibition of Chemical Weapons. This convention is built on, but extends, two earlier documents that failed to provide for effective monitoring or verification: the Protocol for the Prohibition of the Use in War of Asphyxiating, Poisonous or Other Gases, and of Bacteriological Methods of Warfare, signed in Geneva on June 17, 1925, and the Convention on the Prohibition of the Development, Production and Stockpiling of Bacteriological, Biological and Toxin Weapons and on Their Destruction, signed in Washington, London, and Moscow on April 10, 1972. This section summarizes the key provisions bearing on the proposal's verification requirements. If the United States is to have confidence that the terms of the treaty are being complied with by the other signatories, most important, by the USSR, it will have to develop unilateral means and negotiate effective cooperative procedures to monitor compliance with these specific provisions. The technological devices that could be utilized to support these unilateral and cooperative verification procedures are surveyed in the next section.

The basic prohibitions, contained in Article I, state,

Each Party undertakes not to:
(a) develop, produce, otherwise acquire, stockpile, or retain chemical weapons, or transfer chemical weapons to anyone;
(b) conduct other activities in preparation for use of chemical weapons;
(c) use chemical weapons in any armed conflict; or
(d) assist, encourage, or induce, directly, anyone to engage in activities prohibited to Parties under this Convention.

The broad commercial usage of many of the precursor chemicals that also can be utilized to produce CW agents complicates the verification regime of the convention. Therefore, the convention lists specific "permitted purposes" for which designated agents and precursors may be used legitimately. They include "industrial, agricultural, research, medical or other peaceful purposes; protective purposes; and military purposes that do not make use of the chemical action of a toxic chemical to interfere directly with normal functioning of man and animals so as to cause death, temporary incapacitation or permanent damage." "Protective purposes" in this context means purposes directly related to the acquisition of protection against chemical weapons. The retention, production, acquisition, and use of super-toxic lethal chemicals and key precursors for such protective purposes are strictly limited by the convention to those amounts that could be justified for such purposes.[1] At no time, moreover, may the aggregate amount of super-toxic chemicals or precursors exceed one metric ton, nor may the aggregate amount acquired in any calendar year exceed one ton.

The proposed convention further requires that all stocks, production facilities, and past transfers of chemical agents be declared and that parties make an annual declaration of all key precursor and toxic chemical stocks devoted to protective purposes. All stocks and facilities listed on this declaration and not permitted by the convention most be destroyed.

In addition, the convention provides for a consultative committee with the responsibility to

> carry out systematic international on-site verification, through on-site inspection and monitoring with on-site instruments, of:
> (i)   chemical weapons,
> (ii) destruction of chemical weapons,
> (iii) closure and destruction of chemical weapons production facilities,
> (iv) permitted single specialized facilities for production of super-toxic chemicals. . . , and
> (v)   production for permitted purposes. . . .

The parties are required to allow continuous monitoring with on-site instruments and the presence of inspectors during the actual destruction process, and they must agree not to interfere with the conduct of any of the approved monitoring procedures.

---

[1]The super-toxic lethal chemicals are defined in the convention as those "which have been stockpiled as chemical weapons or which pose particular risk of such stockpiling." Other dangerous chemicals are listed that pose a particular risk of diversion to chemical weapons purposes.

Further provision is made for a fact-finding panel to be appointed by the consultative committee; each member of the panel has the right to request at any time a "special on-site inspection" of any party to the convention. Within forty-eight hours of such a request, the party to be inspected is required to provide the inspection team unimpeded access to the designated suspect location or facility. Additionally, each party has the right to request an "ad hoc inspection," which if approved in twenty-four hours by the fact-finding panel, will commence within the next twenty-four hour period.

Some of these provisions of the proposed convention, of course, may be changed substantially or dropped altogether during the negotiations; some countries have taken strong exception to some items. In this brief, however, it is assumed that the final convention will not differ greatly from the current draft and, *particularly*, that on-site monitoring will continue to be considered *absolutely essential* to maintain confidence in compliance with the articles of the convention. In view of the strong stand that the United States has taken on the need for effective verification, this assumption is eminently reasonable.

Summary of Recent Negotiations on Verification Issues

Both bilateral and multilateral negotiations on the CW convention are being carried out in the Conference on Disarmament. Former U.S. Ambassador Donald Lowitz summarized the status of the negotiations at the end of 1986 and identified the major unresolved negotiating issues requiring prompt attention: (1) declaration and monitoring of chemical weapons stockpiles, (2) elimination of chemical weapons production facilities, (3) prevention of the misuse of the chemical industry for chemical weapons production, and (4) challenge inspection.

In addition, several special efforts by individual countries that are participating in the CD were concluded in 1986 and have helped define approaches to overcoming difficulties in verification. First, for example, The Netherlands convened a Workshop on the Verification of a Chemical Weapons Ban on June 4–6, 1986. The primary focus of this workshop was to review the results of an experimental on-site inspection conducted by The Netherlands to verify the nonproduction of chemical weapons in a large chemical production complex and to develop and evaluate procedures for such inspections that should be required by the convention. The company area where the "experimental inspection" was conducted included a variety of petrochemical and multipurpose chemical production facilities, the largest of which was selected for the exercise. The facility is used to produce an organophosphate pesticide for which trimethyl phosphite is one of the precursor chemicals. Trimethyl phosphite also could be a precursor for the reproduction of chemical weapons.

The aim of the experimental inspection was to study and test organizational and technical aspects involved in routine inspection of a chemical

plant under a CW convention.[2] The workshop made these preliminary conclusions: verifying that prohibited agents are not being produced may be
possible at acceptable costs, but no single procedure would suffice for all
plants; inspectors should visit the plants to be made familiar with plant
operations; and inspectors must be highly qualified. The participants in
the experimental inspection from the chemical plants were said to be
cooperative.[3] Although this experiment in no way solves all the problems
of commercial plant inspection, it is an encouraging first step in the study
of this difficult procedure.

Second, the Norwegian delegates to the CD reported the results of
a six-year study on verification of the alleged use of chemical weapons
in a variety of weather conditions, but particularly in snow. The main
thrust of this work was to develop satisfactory sampling procedures so
that integrity of the samples could be maintained and the CW agents could
be analyzed successfully. Third, the delegation from Belgium reported a
study on scheduling the destruction of chemical weapons in accordance
with the proposed convention.

While some progress thus has been made toward overcoming the
difficulties of monitoring treaty compliance, the problem of effective verification continues to be compounded by several considerations:

Availability of Chemicals. Many of the precursor chemicals used in the manufacture of CW agents are produced and transported in very large quantities in many countries for permitted commercial uses (herbicides,
insecticides, pharmaceuticals, and so on).

Availability of Production Facilities. The plants in which CW agents are, or
can be, produced also can be used for the production of many permitted
chemicals, and certain types of permitted commercial plants also could
produce CW agents.

Extent of Monitoring Requirements. There are a great number of permitted
commercial chemical production plants, which would require extensive
monitoring capabilities to ensure that prohibited agents were not being
produced.

Toxicity. Chemical warfare agents obviously are extremely toxic, but so
too are many permitted chemicals. The special industrial prophylactic

---

[2]"Workshop on the Verification of a Chemical Weapons Ban, held in the Netherlands, 4–6
June 1986." (Paper presented by the Netherlands at the Conference on Disarmament, Geneva.
CD CWWP.141, June 10, 1986.)
[3]Australian government representatives, in cooperation with officials of the Australian chemical industry, also conducted and evaluated a trial inspection similar to the study presented
at the Netherlands workshop.

procedures required at plants producing permitted toxic agents would make on-site inspections and monitoring more difficult.

Chemical Instability. Some CW agents are somewhat unstable chemically when deployed in the field, making their detection and unambiguous identification difficult; inspectors would need to be on site within hours to days of an illegal CW use to be able to demonstrate that a violation had occurred.

Lack of Necessary Instrument. The portable, rugged, reliable instrumentation required for effective, practical, on-site monitoring has not yet been developed, so it is not yet known if it will perform adequately in a variety of inspection procedures. Nevertheless, recent developments suggest that significant progress toward verification of treaty compliance could be made in the near future by a coordinated and well-directed effort. Independent accomplishments pertinent to *parts* of this problem have been made in many countries—Canada, Finland, West Germany, The Netherlands, Norway, Sweden, Switzerland, and the United States. Chemical and biochemical techniques have been demonstrated by European and North American researchers to be effective both in detecting and analyzing chemical agents. The Chemical Warfare Review Commission, appointed by President Reagan in March 1985, has focused renewed attention on the unsatisfactory CW posture of the United States, which should add to the determination of government officials to correct deficiencies and to be prepared to monitor a CW treaty effectively.[4]

If the United States is to be prepared to monitor a CW treaty, however, it is essential that the government launch a forceful effort to stimulate additional research and development of verification and monitoring capabilities and to coordinate such measures with the treaty negotiations. The U.S. CW community should be tasked to develop specific procedures and instrumentation that will provide effective monitoring capabilities consistent with all aspects of the current draft convention and with future amended drafts as the negotiations progress. Negotiators and those who develop these procedures and instrumentation should cooperate closely to ensure that the provisions that are ultimately negotiated can be monitored satisfactorily. They must insist that pertinent technologies be used to develop simplified, rugged, accurate, and reliable monitoring instruments whose availability can support and facilitate negotiations. This instrumentation must offer the necessary verification capability without divulging proprietary or sensitive information or providing unwanted transfers of American technology abroad. In addition, the United States

---

[4]Superintendent of Documents, *Report of the Chemical Warfare Review Commission* (Washington, D.C.: GPO, June 1985).

must define the baseline levels of confidence it can expect from each specific monitoring capability. Therefore, any instrumentation developed must be tested against the requirements of realistic monitoring scenarios.

The United States can, through a focused research and development program, devise equipment and testing such that the levels of confidence in both procedures and equipment can be evaluated. By mobilizing technologies now available for CW treaty monitoring, and the required scientific personnel and facilities, the United States can take an initial, viable step toward the goal of effective disarmament in the CW arena. These technologies, and their applications, are described in the next section.

Technologies Potentially Useful in Monitoring a CW Treaty

In an attempt to develop procedural and technical options for compliance monitoring of the proposed CW convention, the U.S. Army Chemical Research, Development and Engineering Center hosted a Chemical Weapons Treaty Compliance Verification Workshop in 1985. The group of experts established monitoring inspection scenarios and assessed the instrumentation and technologies required to conduct effective monitoring in each. Also, rough levels of confidence in monitoring results were projected. As shown below, the group's findings confirm that the United States has the potential means to monitor effectively a CW treaty.

This section first lists those specific items in the proposed CW convention that the army verification workshop determined would require monitoring and inspection. It then surveys the technologies that could be used in carrying out such compliance monitoring. Finally, it explains the possible applications of relevant technologies in specific inspection scenarios. The technology survey includes instruments currently, or soon to become, available from U.S. and foreign research organizations, as well as some that are still being developed. Where appropriate, novel concepts that would require considerable research and development are also mentioned.[5]

Provisions Requiring Monitoring or Inspection
1. Declared chemical agents and precursor stocks
2. Declared agent production facilities
3. Single permitted CW production facility
4. Special on-site inspections (Article X, CW convention)

---

[5]Foolproof tags for containers, physical security networks of pressure, temperature, motion and distance sensors, and video monitors could be helpful in monitoring a variety of arms-control agreements. These types of devices already are fully developed, readily available, and already in use for a variety of purposes. They will not be discussed further in this brief; nor will the administrative and procedural requirements for initiating and conducting inspections, the physical protection required for inspectors, or the decontamination technologies.

61

5. Ad hoc on-site inspections (Article XI, CW convention)
6. Demilitarization of stocks
7. Agent facilities destruction
8. Movement of stocks to demilitarization sites
9. Chemical training exercises for defensive purposes
10. Commercial chemical production
11. Transport of permitted quantities of CW agents
12. Nonpermitted use of chemical weapons (warfare, terrorism, and so on)

Technology Survey
A variety of technologies could be useful potentially in detecting and identifying prohibited chemical agents and in determining possible violations of treaty provisions pertaining to the storage, production, or use of such agents. To be useful, such technology must be adaptable for use in instruments that could be transported quickly to the site of suspected violations of the treaty. Particularly, those instruments that would be used to assess possible violations of the prohibition against uses of chemical weapons, which might well take place in remote parts of the world, would have to be rugged, portable, and self-contained. In most cases, instrumentation would be used to support on-site inspections by personnel of challenging nations or of international organizations established by the treaty. In some cases, however, instrumentation could be used remotely to supplement and, perhaps, reduce requirements for human inspectors.

Colorimetric Test Kits. These kits are designed for fairly quick and easy identification, in the field or laboratory, of specific prohibited chemical agents. The agents react with chemicals in the kits to turn the test solutions specific identifying colors. Some of these kits already have been fully developed and are deployed with U.S. forces (for example, ACADA, U.S. Army M256, M272, M43A1, and M8 kits). Others require substantial research, development, testing, and evaluation before becoming operational.

Simple "dipsticks" on which chemicals that had been absorbed would react with specific chemical agents to provide identifying color reactions also are available. Both the kits and the dipsticks may be useful to detect some CW agents in the field, but they would not provide sufficient qualitative or quantitative information for many inspection requirements, as discussed under "Applications."

Infrared (IR) Spectroscopy. This analytical method can provide a high degree of accuracy in identifying specific chemical compounds for which "library sample" spectra are available. The sample to be analyzed must be reasonably pure (> 95 percent), and a fairly large sample (10–20 mg) would have to be obtained for confident results. To make this method readily usable for verification monitoring, moreover, there must be improvements (ruggedness and miniaturization) in the portability of the instruments, and a computer library of the spectra of specific chemical agents, precursors, decomposition products, and pertinent chemical reactive combinations of atoms must be prepared. This infrared system could be automated and computer controlled for on-site inspections and easy use in the field. Infrared spectrometers can be made smaller and more portable than mass spectrometers.

Fourier-Transform-Infrared (FT-IR) Spectroscopy. This infrared system has the advantage that it can operate effectively with a somewhat smaller and less pure sample than the IR spectrometer, but it probably would not be significantly superior to the computer-adapted system described above. The FT-IR system also would need a spectrum library and would have to be made more portable.

Gas Chromatography (GC). This type of analysis can be used to help determine the quantities of various components in a mixture, as well as to help identify those specific compounds in a mixture for which chromatography data are available. It can also purify samples. Single-channel chromatography cannot be relied on for definitive identification, but there is available from Nordion of Finland a dual-channel, high-performance instrument that would be more reliable than single-channel systems. The Norwegians, for example, have used chromatographic procedure in their field exercises to collect and purify small samples of CW agents. Some chromatography systems now available are portable, but adaptation for use in field inspections would require additional development.[6] A number of useful variants of chromatography also exist, including liquid systems, glass capillary models, and flat plate absorption techniques that have varying capabilities for purifying samples, working with small samples, identifying compounds, and making quantitative analyses.

Tandem Mass Spectroscopy (MS/MS). This analytical system uses two cooperative mass spectrometers and has excellent sensitivity. It can be used to identify picograms (10^{-12}g) of a specific material in the presence of a mixture of impurities. Most current systems are relatively large and require a high level of scientific expertise to obtain good analyses. A system that can be transported in a utility van is available, however, from the Sciex Company of Canada. In addition, development of a more easily portable tandem mass spectrometer has been accomplished by the Bruckner-Franzen

---

[6]Personnel of the Swedish Ministry of Defence, FOA-4 laboratories have developed a small field model that incorporates a photoionization detector.

Company for a system included in a reconnaissance vehicle of the army of the Federal Republic of Germany. This system could be adapted readily for use in CW treaty compliance inspections.

Tandem Gas Chromatography-Mass Spectroscopy (GC-MS). This method combines gas chromatography with mass spectrometry. It can provide excellent identification of very small samples, but it presently is a large and complicated system requiring scientific support and operation. It would require considerable development to adapt this system for verifying compliance with a CW treaty.

Nuclear Magnetic Resonance Spectroscopy (NMR). This analytical method involves observation of nuclear spin components in molecular structures. Spinning of certain parts of an atomic nucleus (for example, the protons), causes an interaction with the magnetic forces of the atomic electrons. When atoms are combined into molecules, components of these spin effects in one atom are changed by forces from neighboring atoms. These component changes can be measured by nuclear magnetic resonance. NMR is used to derive structural features of compounds for precise identification and is often used in conjunction with IR, GC, and MS for accurate descriptions of the detailed structure of chemical compounds. This includes the identification of isotopes substituted into molecules, as could occur to create new chemical agents.

NMR would be useful determining the molecular structure of new or suspected chemical agents. The IR, GC-MS, and NMR instruments all would be invaluable, of course, for the precise identification of samples that could be returned to a central laboratory. Such instruments are used in standard practice for difficult analyses.

Coated Piezoelectric Crystal Array Analyzers. The operating principle of this analytic tool is based on the fact that the vibrational frequency of piezo-electric crystals is shifted a measurable amount as chemical agents are adsorbed on polymer coatings on the crystals. By using an array of crystals with different specific polymer coatings, definitive patterns in the frequency shifts can be obtained. When fully developed, these instruments should have the sensitivity to detect ten parts of chemical agent per billion parts of gas with a five- to twenty-second response time. They can be made small and portable; an experimental model of about attaché-case size has been demonstrated. With computer control and a computer library of relevant chemicals, they could be useful for on-site inspections. The system is being developed now at the Lawrence Livermore National Laboratory. Recent work with surface acoustic waves on piezoelectric crystals at the Prins Maurits Laboratory in The Netherlands shows promise for small sensors with high sensitivity.

Coated Fiber Optrodes. The operation of these devices is based on a polymer-coated fiber and functions in one of two ways. First, the fiber may change its refractive index as specific CW agents are adsorbed by the polymer. Second, the stress resulting from adsorption may bend a fiber that is bonded on one end so that light passing through the fiber (optrode) moves on a measuring scale. These instruments, which are now being developed, can be made as small as a pack of cigarettes. They can be used for the specific identification and some quantitative analysis of CW agents, precursors, and decomposition products. Their small size would potentially make them useful in on-site challenge inspections. Both the polymer-coated crystals and the optrodes also could be useful to the military services as CW warning sensors and for CW stockpile monitoring sensors.

Immunoassay. A breakthrough has occurred in the capability, ease of use, and field compatibility of antibody methods of analysis by combining three separate technological advances: monoclonal antibody technology, fluorescent waveguide immunoassay, and a non-volumetric rapid immunoassay technique. Thus, it is now possible to develop an antibody assay system that combines the extreme specificity of monoclonal antibody techniques with a sensitivity better than picograms (10^{-12}g). These technologies can be used to develop a miniature portable sensing system requiring no perishable fluid reagents, no metering of fluids or any other difficult manipulation under field conditions, and no special operator skill requirements. Work on this system also is in progress at the Lawrence Livermore National Laboratory. This instrumentation would be excellent for on-site inspections in the field, where very small samples of impure milieus are likely to be found.

Enzymatic Assay. An enzymatic assay system could very likely be applied to the CW monitoring problem, although much further development effort would be required. This assay system could provide a measurement of the physiological effects experienced by both humans and animals in the vicinity of CW production, use, or contamination, which could reveal the existence of new agents through *specific* suspected generic responses. It could be useful in on-site inspections where certain types of previously unknown and therefore uncatalogued agents or precursors are suspected. For example, analysis of levels of acetylcholinesterase in a person's blood, or differences between a person's hemoglobin oxygen-binding capability and normal levels, could provide evidence of new toxic agents.

Another generic detection scheme would utilize receptors—the physiological targets of the chemical agents in humans. Detectors using receptors would be able to establish the presence of agents not presently known. This capability would match the intent of the proposed convention to monitor chemicals having adverse effects on humans. This of course, involves monitoring for more than the classical warfare agents.

Association of Selected Instrumentation with Data Links. Unmanned instruments potentially could monitor the destruction of chemical stocks or the production of agents for permitted purposes. Probably any of the instruments and devices described above could be adapted for unmanned use, but the mass spectrometers and NMR would be the most difficult to adapt. Tamper-resistant data links would have to be provided to give a prompt alert of possible interference with any unmanned instruments or with the data transmission itself. Appropriate technology appears to be available, but it would need to be adapted to this purpose and tested under simulated field conditions. This could be an expensive project, but it would be an important factor in reducing requirements for a human presence for inspection purposes. There will still be a strong requirement for inspection visits to ensure nontampering with on-site devices and to allow calibration, repair, testing, maintenance, and other tasks of the devices.

Applications

Several of the convention monitoring requirements can be accomplished by using a similar set of devices or instruments. In the following discussion, monitoring scenarios with similar monitoring or inspection requirements are grouped together.

Declaration of Existing Stocks and Production Facilities. The proposed CW convention currently calls for a declaration by each party, within thirty days after the convention enters into force for that party, of all existing chemical weapons, chemical weapons production facilities, and super-toxic lethal chemicals or key precursors for protective purposes and their production facilities. It further requires a general plan for the destruction of all chemical weapons and chemical weapons production facilities, except for a single permitted production facility and chemical weapons allowed for specific purposes.

It appears that the chemical aspects of monitoring and verifying the declaration of chemical agents and precursor stocks, the movement of stocks to demilitarization sites, and the transport of permitted quantities of CW agents (items 1, 8, and 11 on the earlier list) could most likely be accomplished by using simple detection and analysis test devices and instrumentation. Because the chemical agents or precursors in question would be previously identified by the party being inspected, it would appear that verification of the specific identity (and quantity) of the declared items could be made simply with the U.S. Army colorimetric test kits, ACADA, or other technologies, such as dipsticks, coated fiber optrodes, IR spectroscopy, mass spectrometry, or the coated piezoelectric crystal analyzer. The feasibility of most of these methods has been demonstrated; however, additional efforts to develop the hardware and to improve the portability of some of these instruments and devices are needed. Procedures for the use of such monitoring equipment on site must be carefully planned to ensure the highest possible level of confidence in the results as compared to the monitoring requirements.

For verifying declarations of agent production facilities, the single permitted CW production facility, and destruction of agent facilities (items 2, 3, and 7), again, simple chemical procedures and equipment should be sufficient. The specific agents to be verified would be declared, greatly simplifying necessary tests to identify the specified chemicals as agents and precursors. The instrumentation mentioned in the previous paragraph would apply to these activities as well, although they may not be as important as the administrative and reporting procedures required to ascertain full compliance.

On-site Inspections. More difficult issues are implicit in the requirements to carry out special on-site inspections and ad hoc on-site inspections (items 4 and 5). In these situations there would be a presumption of non-compliance, and it may be assumed that the host party would be uncooperative. Also, it is possible that some previously unspecified or unknown precursor, CW agent, chemical process, or decomposition product might be involved. The instrumentation required thus would have to be more sophisticated technically than that used to verify the destruction of specific types of declared chemical agents, and yet it must still be workable under adverse circumstances—in the field, on short time scales, with smaller and less pure samples, and perhaps in the face of deliberate efforts to deceive, interfere, or intimidate the inspectors.

In addition to the type of detection and analysis equipment and instrumentation just mentioned, it would be useful in carrying out on-site inspections to have portable (by personnel or small vehicle) models of gas chromatographs available, as well as tandem mass spectrometers, FT-IR and NMR spectrometers, immunoassay, and glass capillary chromatographs. Not all of these instruments or methods would be required in each case, but all could be useful in particularly difficult analyses. Considerable development effort will be required to bring some of these instruments to a status at which their use would be feasible in ad hoc or special on-site inspections.

Monitoring the Destruction of Existing Stocks and the Operation of the Single Permitted Production Facility. Monitoring the demilitarization of CW stocks (item 6) and, to some extent, the operation of the single permitted CW production facility (item 3) pose different problems, since the necessary operations may span many years. The draft convention requires that the destruction of chemical weapons begin "not later than twelve months, and finishing not later than ten years, after the Convention enters into force" for each party. To monitor compliance with these aspects of the draft convention, the use of unmanned instruments to monitor on-line processes, including measurement of their pressure, volume, temperature, chemical analysis, electrical loads, natural gas flow, as well as to analyze decomposition products, would be imperative, so that continuous monitoring by human inspectors might not be needed.

Apparently, data from the instrumentation would need to be forwarded to the consultative committee in secure form and in a manner to indicate that no tampering had occurred. Alternatively, the instruments could be programmed so that an alarm would be given via a secure data link to the inspection force of the committee if an indication of noncompliance or tampering were detected. These procedures are yet to be determined. Special tamper-resistant equipment would need to be developed for this purpose and for other scenarios requiring on-site equipment. Computer-controlled process monitors, perhaps using remotely terminated process sensors providing data to a central automated laboratory, should be satisfactory for some purposes.

Some of these monitoring and analysis instruments are currently available, as are some aspects of the computer control and data communication subsystems. Development of an optrode system (fiber optics used in chemical and process analysis) now appears feasible, and it could greatly reduce the cost of monitoring the many existing CW sites. A special effort must be included in this development to uncover any inadequacies and to evaluate the likely limits of confidence in the system. The more sophisticated instrumentation called for in connection with ad hoc on-site inspections may not be required for monitoring the destruction of stocks and permitted production, but the monitoring protocol and general inspection procedures required for the latter may be especially difficult to negotiate and carry out.

Monitoring Permitted Exercises and Commercial Chemical Production. Monitoring defensive chemical training exercises and commercial chemical production (items 9 and 10) probably will pose difficult problems of general inspection procedure. The instrumentation proposed for previous inspection provisions could be used to monitor the chemical aspects of these two operations, but negotiation and implementation of an effective verification protocol may be extremely difficult, and the level of confidence in the monitoring results is uncertain at present.

Monitoring for the Nonpermitted Uses of Chemical Weapons. Monitoring nonpermitted uses of chemical weapons (item 12) may be the most difficult task of all, as has been demonstrated to some degree by U.S. attempts to demonstrate conclusively CW deployments in Afghanistan by the Soviet Union and the uses of toxins in Southeast Asia by Vietnam. Validating allegations that Iraq used chemical agents, on the other hand, has proved far easier, primarily because the aggrieved party, Iran was in a position to grant rapid access to the site of the alleged use and cooperated with investigators in other ways as well.

In verifying allegations of the use of prohibited chemical agents, all the chemical detection and analysis equipment and instrumentation proposed for use in on-site inspections, if made sufficiently portable, should be helpful—including, especially, the immunoassay and enzymatic systems. The medical examination and the biochemical or clinical analysis of alleged victims of chemical warfare also may be useful. The physiological examination of animals, birds, and perhaps plants in the suspect area could be useful as well, but on-site visits would still be imperative. Special problems of sample collection and sample security may add significantly to the verification effort. Rapid access to sites of suspected use is critical in reaching conclusions as to culpability.

Potential Advantages of Remote Sensors

The previous section described how both on-site and remotely positioned sensors could aid in the verification of a chemical weapons treaty. The degree to which remote, unmanned sensors might limit the number and duration of the required on-site human inspections, as well as reduce the need for a continuous human presence during the demilitarization of prohibited chemical weapons stocks and the permitted production of chemical warfare agents, is an important question. Although the technologies for a network of temperature, pressure, motion, and distance sensors, as well as video monitors, are available now, research and planning would be required to design site-specific, tamper-resistant networks.

A remote sensing system could be based, for example, on chemical flow monitors connected to a central analytical instrumentation area, where the analysis schedule and the data reporting would be controlled by computers. Combined with computer-monitored and -controlled flow and process sensors (now available), such a system could provide, through secure data links, either continuous or periodic analytic results or a warning signal if production or destruction processes surpassed previously established tolerance boundaries.

Additionally, sensors with high sensitivity (that is, down to picogram levels), based on coated piezoelectric crystal arrays or the coated fiber-optic instruments, could be used in remote sensing systems. These might be emplaced outside the process areas but within specific buildings, or on or near the external walls of buildings, or possibly on the perimeters of chemical weapons research, production, or storage areas. These sophisticated sensors could be connected to data links (possibly land lines, radio, or satellite) for direct communication with the consultative committee facilities. They obviously would need to be tamper-resistant systems. In some cases of materials with moderate- to high-vapor pressure, they may provide real-time warning of the presence of disallowed chemical agents, precursors, or decomposition products of known or suspected agents. It is probable that an immunoassay instrument using appropriate biomaterials also could be developed and incorporated into such a detection and warning system to report the presence of specific chemical groups indicating the presence of certain types of chemical warfare agents. Research would be time-consuming, however, and would require an interdisciplinary team.

All such remote sensing systems still require development and would need to be adapted to each facility being used to demilitarize CW agents or to carry out permitted production. If such development programs were successful, however, and provisions governing the installation of remote sensors successfully negotiated, then requirements for more intrusive human inspections might be reduced and the overall negotiability of a CW treaty improved.

Thus, it appears feasible to develop remotely emplaced sensors from known technologies for monitoring the destruction of declared stocks and permitted production, to warn of the presence of CW agents in unauthorized locations, or possibly to detect some new agents. Much development would be required, however, to increase the ruggedness of, and particularly to miniaturize, these instruments, and also to tailor any single system to a particular site. In addition, it would be necessary to conduct extensive testing and analysis to obtain estimates of the confidence levels obtainable with these systems. Remote sensors should reduce the requirements for on-site inspections and a human presence; however, they would not eliminate these requirements.

Possible Spin-Offs from the Technology Developed for CW Verification
The new miniaturized instrumentation and devices that might be developed for CW verification monitoring could be useful in other applications. Some, for example, could be employed with little, if any, modification in monitoring U.S. production, stockpiling, and, particularly, the destruction or demilitarization of chemical weapons now obsolete, whether or not a treaty were negotiated. The United States has useful instrumentation now for these purposes, but some novel concepts could be realized from a coordinated research program that would enhance these capabilities. With some minor modifications in software (libraries) or sampling techniques, this new instrumentation also would be useful for safety monitoring in research and production of many hazardous chemicals and other materials used for civilian purposes, although there are sophisticated systems now in use commercially from which the CW community has obtained valuable concepts and insights.

Second, an *advanced* flow and process control system using optrodes with one central computer-controlled analytical instrument and communication system, as envisioned for the remote monitoring of the destruction of existing stocks of CW agents, could be adapted to chemical production plants with a high probability of better flow and process control of both local and remote operations at a significant reduction in cost.

Third, the military services could benefit from advances in piezo-electric crystal array and coated fiber-optic technologies, as they might be used for the detection and rapid analysis of chemical warfare agents and other noxious gases on the battlefield. They might also be applied as warning sensors on ships and aircraft for the detection of noxious exhaust or other gases.

Fourth, the generic agent detectors could have applications for the general detection of classes of harmful substances or poisons in the atmosphere or other environments.

Risks and Liabilities

It is apparent that the monitoring of a chemical weapons convention, if one can be negotiated, will be a difficult, expensive, and time-consuming operation. Major aspects of the problem are the ubiquitous production and use of chemicals and the facts that both a large number of commercial plants could be diverted relatively easily to CW agent production and that the precursors for chemical weapons have many innocuous and commercial uses. Another crucial factor is that new, lethal chemical agents could be developed fairly readily, so that detection and analysis equipment must be versatile and sophisticated. It would be important to have generic detectors that could monitor the action of new agents on the physiological receptors in humans.

There are important additional concerns. The design of instruments used to monitor a CW agreement must not include critical technologies that the United States wishes to keep secret, as many of these instruments may have to be provided to the consultative committee for examination or use. The Soviet Union and other countries will almost certainly insist on obtaining copies of any instruments utilized on their soil. In addition, the monitoring process must not disclose chemical product, process, or manufacturing technologies; this a serious concern of commercial firms.

Another aspect of verification monitoring of special concern is that challenge or ad hoc on-site inspections could be used by foreign governments as a means to gain access to sensitive U.S. government facilities—for example, nuclear weapons laboratories or production and storage facilities. Because inspection procedures proposed to be used by the United States at Soviet sites also would be used by the Soviets as U.S. sites, U.S. policymakers have to consider duly the effects of this double-edged sword.

On-site or remote sensors that are unattended will need to be cleverly protected to avoid deception and tampering. Although substantive improvements have been made in tamper-resistant systems for nuclear weapons, further work would be required to apply these principles to remote sensors to monitor chemical weapons development, production, and stockpiling activities.

Some have argued that binary chemical weapons, such as those now ready for production in the United States, may make it more difficult to verify compliance with a CS treaty, because no super-toxic chemical agents need to be produced or stockpiled. If other countries produce binaries or certain other novel weapons (wherein more than just the final chemical reaction is completed after weapons launch), this argument continues, verification concerns could be compounded. On the other hand, the Chemical Warfare Review Commission concluded that the U.S. production of binary chemical weapons would not make the negotiation of a "multilateral, verifiable ban on chemical weapons" more difficult. In fact, the commission maintained, the U.S. decision to produce binary weapons may even have helped bring the Soviets to the Geneva negotiations with a more positive attitude.

Conclusion

This brief survey of technologies and instruments potentially relevant for verifying compliance with the U.S. draft Convention on the Prohibition of Chemical Weapons makes clear that there exist a number of ways to improve existing capabilities to detect and identify lethal chemical agents whose production, storage, or use might be prohibited by a CW treaty. If additional research and development demonstrated that these technologies could be incorporated in other instruments with appropriate characteristics, the new technologies could be used to help inspectors verify compliance with any CW treaty concluded and, in some cases, to reduce requirements for the frequency or duration of human inspections. To serve such purposes, development programs must result in highly reliable, portable, and rugged instruments that can be adapted for use in a variety of inspection scenarios, some envisioning their dispatch on very short notice to remote regions of the globe. The greater the flexibility of such instruments in terms of dealing successfully with the range of existing and potential lethal agents, the better. With respect to technologies that might be used in remote (unmanned) sensors, development programs also must ensure that instruments are tamper-resistant and, to the degree possible, that they minimize the degree of intrusiveness required to operate successfully.

The probability that such effective instruments will be developed could be greatly enhanced if the U.S. government were to pursue a coordinated program toward these ends and if it were willing to invest greater research funds toward such an objective. Plans for such a development program have been drawn up by the U.S. Army Chemical Research, Development and Engineering Center in Aberdeen, Maryland, and have already been coordinated with other relevant government agencies. To date, however, the program has not been given a high enough priority to compete successfully for limited budgetary resources.

In this writer's view, the development of inspection instrumentation and its integration in planning for the verification of possible CW treaties should receive a higher priority. Development of such instruments will require a substantial effort over a sustained period. Critical elements should be in hand before the United States goes very far in proposing the details of inspection procedures to other nations. The initiation of a focused and coordinated research program should receive a higher priority in support of the United States' objectives in the Geneva negotiations.

A broader issue also is suggested here. The U.S. government's inability or unwillingness to allocate funds for the development of improved means of verifying chemical weapons treaties is replicated in other arms-control negotiations. One exception is negotiations on limitations on nuclear weapons testing; total government spending on verifying nuclear test bans, including satellite monitoring and the development of seismic monitoring methods, has already reached a level of about $100 million per year. Still, this is the only exception. It seems reasonable that comparable expenditures should be devoted to the development of means to verify limits on chemical weapons, nuclear delivery systems, and the other weapons for which the United States is already engaged in serious negotiations.

The need for a long-term and more substantial commitment to the development of improved means of verifying arms-control agreements also suggests the need to centralize such activities in a specific government agency. Although the actual research and development programs should be carried out in a wide number of appropriate centers, centralization of responsibility for overall technical direction and management could help promote greater attention to the problem as well as ensure a more efficient and effective development effort. It also could facilitate closer coordination between verification research and technological opportunities and the positions actually taken in the various negotiating forums.

Research and development for verifying chemical weapons treaties should be one element of a national verification program. The current inadequacies in the CW situation furnish a convincing rationale for establishing a broader national effort.

——|5|——

U.S. GOVERNMENT ORGANIZATION FOR ARMS CONTROL VERIFICATION AND COMPLIANCE

BY MICHAEL KREPON

Under the National Security Act of 1947, the president of the United States has considerable latitude in organizing the executive branch of government for national security affairs. This act establishes a National Security Council (NSC), but it makes no mention of the president's assistant for national security affairs and the NSC staff. The act further seeks to "enable the military services and the other departments and agencies of the government to cooperate more effectively in matters involving the national security," but does not mandate procedures for accomplishing this objective.

Since this legislation was enacted, presidents have chosen various methods to provide themselves with advice and policy options from within the executive branch. This practice is only natural, because presidents have different management styles and priorities and policy advisers serve at the "pleasure of the president."

At the outset of the Carter and Reagan administrations, the options generated for presidential review (and the bureaucratic disputes generated in the process) were not foreordained by organizational charts. Policy differences among the presidents' senior advisers were subsumed beneath broad campaign pledges regarding arms control. Because these presidential arms control objectives were strongly held, different bureaucratic mechanisms for handling verification and compliance issues probably would not have resulted in different policy decisions. Political appointees and civil servants working on verification and compliance problems

This essay was first published in Michael Krepon and Mary Umberger, eds., *Verification and Compliance: A Problem-Solving Approach* (London and New York: Macmillan Press, 1988). The author is grateful for the comments of executive branch officials who provided background information on a not-for-attribution basis. In addition, Victor Alessi, Jim Blackwell, Barry Blechman, Harold Brown, Robert Buchheim, Dan Caldwell, William Colby, Robert Einhorn, Alton Frye, Rose Gottemoeller, Sidney Graybeal, Gerald Johnston, Mark Lowenthal, Ray McCrory, Michael May, Milo D. Nordyke, Ivan C. Oelrich, John Rhinelander, Gerard Smith, Howard Stoertz and Herbert York provided helpful comments and criticism. Their assistance should not imply support for the analysis and conclusions presented here.

may strive for objectivity, but assessments are often shaped by an administration's philosophy as well as by personal and institutional biases. Ronald Reagan, for example, was first elected president after strongly criticizing his predecessors' efforts on strategic arms control. It is therefore not surprising that interagency deliberations within the Reagan administration did not dwell on options to affirm the objectives and purposes of the prior Strategic Arms Limitation Talks (SALT). Likewise, because Jimmy Carter was supportive of the SALT process when he was elected president, officials within the executive branch during this administration did not seek to implement a policy of dismantling the SALT accords.

Political objectives can change within the time span of an administration, however. When this is the case, presidents are well served when internal deliberations and policy options take this eventuality into account. Presidents are also well served when bureaucratic differences are presented fairly and fully in internal deliberations and when these disputes are not aired in public.

Every president also has a stake in bureaucratic processes that generate policy options in a timely manner. It is extremely difficult for the president and his senior advisers to anticipate the course of negotiations, given all of their other pressing responsibilities. Nevertheless, if they fail to do so, they risk losing the negotiating initiative and the result can be defensive, reactive policies.

For all of these reasons, the organization of the executive branch that addresses verification and compliance issues can make a difference. As the Tower Commission's report concluded, "Process will not always produce brilliant ideas, but history suggests it can at least help prevent bad ideas from becoming Presidential policy."[1]

This essay first describes and then evaluates the organizational structures and bureaucratic approaches used during the Nixon, Ford, Carter, and Reagan administrations to address verification and compliance of strategic and intermediate-range nuclear arms limitations. The Nixon, Ford, and Carter administrations' bureaucratic structures for handling SALT verification and compliance issues will be treated together, as they had much in common despite minor differences in approach and nomenclature. The Reagan administration's approach has been different than its predecessors' and will be analyzed separately. Finally, recommendations will be offered based on the author's observations of the strengths and weaknesses of prior bureaucratic approaches.

---

[1] "Report of the President's Special Review Board," U.S. President's Special Review Board, Washington, D.C., February 26, 1987, V–2.

Executive Branch Organization for SALT Verification

The Nixon administration's procedures for handling SALT verification issues have been well documented in authoritative appraisals and
in memoirs of the American and Soviet negotiators.[2] In 1969 the Arms
Control and Disarmament Agency (ACDA) was entrusted with the authority
to carry out an interagency canvass of broad negotiating options, but it
soon encountered the Pentagon's skepticism about the "verifiability" of
a number of the proposals favored by SALT enthusiasts. Gerard Smith,
ACDA's director, then proposed that National Security Adviser Henry Kissinger establish an NSC-based forum to assess these disputes. The verification panel was thus created in July 1969. It quickly assumed a much
broader mandate that covered a wide range of SALT policy issues. Kis-
ςinger chaired the verification panel. Other members included the deputy
secretary of defense, the chairman of the Joint Chiefs of Staff, the under
secretary of state for political affairs, the director of the Central Intelligence Agency, the director of ACDA, and, during the Nixon administration, the attorney general. As is customary, working groups were also
established, comprised of senior subordinates to the principals involved.
The verification panel working groups were also chaired by an NSC official and supported by interagency study groups under the chairmanship
of the Department of State, Defense, or ACDA.

According to Kissinger's memoirs, concerns over verification were
woven into the verification panel's initial deliberations of SALT negotiating
strategies: "The CIA was asked to assess the verifiability of each weapon

---

[2]See, for example, *U.S. Commission on the Organization of the Government for the Conduct of Foreign
Policy* (Washington, D.C.: GPO, June 1975), Part V, 325–44, hereafter referred to as the Murphy
Commission Report; Mark Lowenthal, SALT Verification, Report 78–142F, Congressional
Research Service, April 24, 1979; Thomas Wolfe, *The Salt Experience* (Cambridge, Massachusetts:
Ballinger Publishing Company, 1979); U.S. Congress, House Select Committee on Intelli-
gence, "U.S. Intelligence Agencies and Activities: Risks and Control of Foreign Intelligence,"
Part V, Hearings (Washington, D.C.: GPO, 1976), hereafter referred to as the Pike Commit-
tee Hearings; Committee on Foreign Affairs and Subcommittee on International Security
and Scientific Affairs, *Strategic Arms Limitation Talks: Hearings and Briefings* (Washington, D.C.:
GPO, 1979), hereafter referred to as the Strategic Arms Limitation Hearings; Raymond L.
Garthoff, *Détente and Confrontation: American-Soviet Relations from Nixon to Reagan* (Washington, D.C.: The Brookings Institution, 1985); Henry Kissinger, *Years of Upheaval* (Boston: Little, Brown and Company, 1982); Henry Kissinger, *White House Years* (Boston: Little, Brown
and Company, 1979); John Newhouse, *Cold Dawn: The Story of SALT* (New York: Holt, Rine-
hart, and Winston, 1973); Gerard Smith, *Doubletalk* (New York: Doubleday and Company,
Inc., 1980); Richard M. Nixon, *RN: The Memoirs of Richard Nixon* (New York: Grosset and Dun-
lap, 1978); Gerald R. Ford, *A Time to Heal: The Autobiography of Gerald R. Ford* (New York: Harper
& Row, 1979); Jimmy Carter, *Keeping Faith: Memoirs of a President* (New York: Bantam Books,
1982); U. Alexis Johnson and Jef Olivarius McAllister, *The Right Hand of Power* (Englewood
Cliffs, New Jersey: Prentice Hall, Inc., 1984); Rose E. Gottemoeller, "Evolution of the U.S.
Organizational Setup for Dealing with SALT," P–6197 (Santa Monica, California: The RAND
Corporation, November, 1978); Alton Frye, "U.S. Decision-Making for SALT," in *SALT, Agree-
ments and Beyond,* Mason Willrich and John B. Rhinelander, eds. (New York: The Free Press,
1974), 66–101.

limitation proposed—how we could check up on compliance; how much cheating could take place before discovery, and the strategic consequences of potential violations."[3] This integrative approach led, early on, to the Nixon administration's decision to focus on ballistic missile launcher limitations instead of constraints on missile inventories; the former were deemed monitorable with high confidence, while the latter were not. President Nixon used the term "adequate" verification to characterize his negotiating objective.

The verification panel's formal role in this and other SALT verification decisions is, however, unclear. Richard Nixon and Kissinger maintained an extraordinarily centralized control over SALT decisionmaking. Formal options, including decisions related to verification, were generally avoided in verification panel meetings. Gerard Smith found that often these sessions were "perfunctory and made little contribution to solving problems, but rather were recitals of departmental positions fairly well known to all hands before the meeting."[4] Nevertheless, the prospect of a series of interagency meetings under the taxing chairmanship of the president's national security adviser forced the relevant bureaucracies to do their homework on verification issues and helped top political appointees to familiarize themselves with these obscure topics. On the most critical SALT decision made by the Nixon administration—whether or not to foreclose the flight-testing of multiple independently targeted reentry vehicles (MIRVs)—verification analysis took a back seat to political expediency: monitoring concerns were used as a false excuse not to effectively block MIRVed missile deployments by means of a ban of flight tests, an action that would have met with strenuous opposition from the Pentagon and its powerful supporters on Capitol Hill.[5]

After deliberations by the verification panel, President Nixon, working closely with National Security Adviser Kissinger and his staff, would select negotiating positions. Gerard Smith's multiple responsibilities as director of ACDA (making him a member of the verification panel as well as a statutory adviser to the secretary of state on arms control matters) and head of the U.S. SALT delegation offered the possibility of close coordination between verification assessments and U.S. negotiating strategy. In practice, coordination was ragged, primarily because of the Nixon and Kissinger obsession with secrecy and control of the bureaucratic apparatus.

---

[3]Henry Kissinger, *White House Years*, 148.

[4]Gerard Smith, *Doubletalk*, 111.

[5]Alton Frye, *A Responsible Congress: The Politics of National Security* (New York: McGraw-Hill, 1975), 47–97; Henry Kissinger, *White House Years*, 212, 540–49; Murphy Commission Report, Part V, 336. For lengthier treatments of the MIRV issue, see Ted Greenwood, *Making the MIRV: A Study of Defense Decision-Making* (Cambridge, Massachusetts: Ballinger Publishing Co., 1975); and Ronald L. Tammen, *MIRV and the Arms Race: An Interpretation of Defense Strategy* (New York: Praeger Publishers, 1973).

The dysfunctions created by back-channel negotiations did not, however, affect verification issues greatly, given the prior negotiating decision to restrict the scope of agreements to units of account most amenable to U.S. monitoring methods. The SALT I end game, therefore, was not over politically sensitive verification issues.

President Gerald Ford utilized the same SALT bureaucratic mechanisms as President Nixon, although SALT decisionmaking became more formalized within the National Security Council. He also adopted the Nixon administration's standard of "adequate" verification. Prior efforts to avoid the rigid association of bureaucracies with negotiating options gave way to serious disputes among agencies, ultimately ending President Ford's attempt to secure a SALT II treaty.

The organizational structures established during the Nixon years were carried over essentially intact to the Carter administration, although given different names. The old verification panel was upgraded to become a cabinet-level special coordinating committee, chaired by the president's National Security Adviser Zbigniew Brzezinski. One of his deputies also chaired an interagency SALT working group that analyzed issues in depth, as directed by their superiors. The views of the intelligence community had particular weight on SALT verification issues, although they were not always decisive. Coordination between Washington and the delegation in the field was improved over that of the Nixon years, as one would expect with the more collegial system preferred by President Carter. But interagency disputes—including differences of view over verification provisions—were common, with agencies advancing their favored positions for decision by the president. These disputes were manageable, however, because the president's senior advisers did not use the verification issue as a surrogate to block a SALT treaty.

SALT II, like the Interim Agreement, sidestepped difficult verification issues by maintaining missile launchers as the chosen unit of account and by expanding coverage to include specific types of heavy bombers, which could be monitored without great difficulty. The problems posed by cruise and mobile missiles were almost entirely postponed to subsequent negotiations. Even so, SALT critics raised the verification issue in their bill of particulars against the treaty, especially its qualitative limits. Two of the provisions that critics found so difficult to monitor—the treaty text governing encryption and new missile "types"—were among the last to be resolved in the SALT II negotiations. Differences in interpretation between the two sides remained in both instances. In the case of the provision defining missile throw-weight, these differences were not explored deeply. In contrast, the two sides grasped the extent to which the treaty text glossed over their differences over the encryption provision. In both instances, administration officials understood the potential verification problems involved. President Carter consciously chose to subordinate verification issues to other priorities—securing a negotiated agreement that would provide a basis to address subsequent concerns over Soviet concealment practices and limiting the growth potential for new Soviet ballistic missiles.

Indeed, verification issues were addressed in a far more systematic way during the SALT II negotiations than previously, largely because of the intelligence community's greater familiarity with both the monitoring and the negotiating process. In contrast to the development of the SALT I accords, a community-wide assessment was kept to track negotiating provisions and U.S. monitoring capabilities. This highly-classified Interagency Intelligence Memorandum became the basis for responding to inquiries from Congress concerning the "verifiability" of the SALT II treaty and for the preparation of less-highly classified analyses prepared by officials from the Office of the Secretary of Defense (OSD) and ACDA. These assessments evaluated the U.S. ability to monitor Soviet compliance with then-current and projected U.S. assets against then-current and altered Soviet military practices. U.S. hedges in the event of Soviet noncompliance were also addressed, as well as the likely consequences of Soviet cheating.

The Carter administration concluded that undetected Soviet noncompliance would not be militarily significant and that the United States could detect cheating of military consequence in sufficient time to take appropriate countermeasures—the classic definition of "adequate" verification employed by the Nixon administration at the outset of the SALT negotiations. As this judgment went well beyond assessments of U.S. monitoring capabilities, it was offered by those responsible for SALT policy and not by intelligence community officials, who declined to play a role in its formulation. Nevertheless, the Carter administration's conclusions had to track with the monitoring assessments contained in the Interagency Intelligence Memorandum, otherwise the administration's credibility in defending the treaty would have been further undermined. For its part, the intelligence community took pains to separate its monitoring judgments from the verification pronouncements of policymakers.

Executive Branch Organization for Verification in the Reagan Administration

In contrast to the Nixon, Ford, and Carter presidencies, the Reagan administration established a decentralized bureaucratic structure for dealing with verification issues. The role of the NSC staff, so prominent in prior years, was reduced dramatically; verification assessments instead were delegated to several different, but overlapping, interagency groups chaired or cochaired by the bureaucracies most concerned. The resulting organization chart resembled a Jackson Pollack painting. The proliferation of analytical channels offered numerous points of entry into the decisionmaking process for all concerned and the near-certainty of bureaucratic delay, if, as in the case of the Reagan administration, the views of participating agencies reflected particularly sharp conflicts.

Administration officials called for a more stringent verification standard for new arms control accords negotiated under their auspices, using the term "effective" verification to distinguish their standard from the previous ones. An Arms Control Verification Committee (ACVC) was established in 1982 by presidential directive to assess the verifiability of treaty provisions for all arms control negotiations, review the monitoring systems required for treaty verification, and analyze Soviet compliance with existing agreements.

The ACVC was also supposed to critique verification proposals developed by interdepartmental groups (IGs) prior to their consideration by the National Security Council. This cross-checking function could not be carried out, however, because of time constraints and because the ACVC and IG principals were mostly the same individuals. ACVC assessments of alternative treaty provisions were seldom possible prior to consideration by the NSC, and its task of reviewing the monitoring systems for verifying various treaties also remained unfulfilled. In practice, the ACVC dealt almost exclusively with compliance issues; with few exceptions, the responsibility for assessing verification questions devolved to a number of interdepartmental groups.

Two IGs were initially created to deal with verification issues in the Strategic Arms Reduction Talks (START) and negotiations on intermediate-range nuclear forces (INF), reflecting the division of labor in the Geneva talks. Both interdepartmental groups were formally cochaired by the assistant secretary of state for politico-military affairs and the assistant secretary of defense for international security policy, although the State Department's representative usually ran these meetings. In 1985, an interdepartmental group was established for the new defense and space talks, chaired by ACDA. All three IGs had a great deal of overlapping membership. The level of representation varied, with the principals involved often replaced by their deputies or senior staff assistants. Working groups for each IG were also created.

Although START and INF issues had been negotiated in separate forums, their verification requirements were similar, at least in the early phases of both negotiations. The interdepartmental groups for these two negotiations therefore established in 1982 the Consolidated Verification Group (CVG) to analyze common issues and to develop measures to verify treaty provisions. The CVG, like the interdepartmental groups that spawned it, was cochaired by representatives of the Departments of State and Defense. In 1986, the Ad Hoc Group on INF Verification was formed to draft a series of decision papers for consideration by the NSC. The Ad Hoc Group was chaired by the principal deputy assistant secretary of state for politico-military affairs. Over time, the distinction between the CVG

81

and the Ad Hoc Group disappeared because of their common member-ship. Meanwhile, the IGs continued to meet on START and INF issues.

Delay was endemic to the Reagan administration's deliberations on verification for many reasons. The task of devising verification provisions was time-consuming because the issues were complex. Added delays were caused by changes in U.S. and Soviet negotiating proposals that had sig-nificant impacts on perceived verification requirements. These reasons for delay in addressing verification issues were reinforced by the Reagan administration's patterns of bureaucratic politics and procedure. As one centrally involved official noted in an interview, "There was a great reluc-tance by senior policymakers to even discuss verification issues due to the complexity of the issues, the politically controversial nature of the decisions, and a preference to avoid the fact that a number of our militarily significant proposals were going to pose serious verification problems."

The work of the IGs and the CVG suffered because of the differences of view among the agencies involved and the inability of their relatively low-ranking chairmen to move the process forward in the absence of higher-level agreement to do so. Because the NSC staff was not directing the process, it had no "tasking" authority to force options and issue papers in a timely manner from interagency groups. Nevertheless, this process appeared to serve one aspect of President Reagan's management style—a strong aversion to resolving issues when important differences existed in interagency deliberations unless outside events, such as Soviet negotiat-ing tactics, forced the administration's hand.

Highly technical analyses produced by relatively low-level interagency groups were also pursued in too much isolation from the broader policy issues involved. Separate verification and negotiating policy issue papers would work their way through bureaucratic channels, usually addressing each other's concerns in no more than a glancing fashion. Thus, in the INF negotiations, President Reagan approved sweeping, technical proposals for on-site inspections (OSI) produced by interagency committees only to re-verse himself when the primary focus of interagency concern switched from the technical to the policy issues involved. The administration's de-cision to scale back its verification proposals was tied to the Kremlin's decision to accept a global ban on intermediate- and shorter-range nuclear missiles. Nevertheless, monitoring problems—such as determining the ac-tual number of nondeployed missiles in Soviet inventories—that provided at least some of the initial impetus for the Reagan administration's strin-gent verification proposals remained after the Kremlin accepted the zero option.

The interdepartmental groups were supposed to report their recom-mendations to higher-level decisionmaking bodies, which at various times in the Reagan administration have been known as the Senior Interdepart-mental Group, Senior Arms Control Group, and the National Security

Planning Group. For four years, verification issues did not figure promi-
nently in these deliberations because of bureaucratic disputes within the
administration and because top-down authority to force verification issues
upward for decision was lacking. The Kremlin reinforced the Reagan administration's tendency to let verification issues slide in the negotiations
by adopting a position that such detailed discussions should await the resolution of more central issues. Thus, until 1986, verification continued to
be what one administration official described as "the virgin playing field
for the negotiating endgame"—both within the U.S. government and with
the Soviets.

The delay in dealing with verification issues also reflected a conscious
policy choice by Reagan administration officials who believed that their
predecessors had made a crucial mistake in focusing efforts on what was
"verifiable" instead of what was strategically significant. President Reagan's
advisers aimed instead at first defining reductions of military consequence
and then investigating the verification measures required to monitor them
successfully.

To begin with, these criteria were applied irregularly. In the START
negotiations, the administration proposed throw-weight and total missile
inventory limitations—two measurements of military significance—while
postponing the difficult verification considerations associated with reductions in these units of account. The Reagan administration pursued a
different tack for sea-launched cruise missiles (SLCMs), initially refusing
to engage in negotiations because of the a priori assumption that SLCM
limits were not verifiable. Nevertheless, the administration proceeded to
flight-test and deploy SLCMs, presumably a reflection of their strategic
value. To confuse matters further, the Reagan administration proposed
a complete ban on chemical weapons—an agreement that would be at
least as difficult to monitor as controls on SLCMs.

Leaving these inconsistencies aside, the difficulties involved in hold-
ing verification issues for the last phase of a negotiation, instead of integrating them at the outset of an administration's deliberations over
negotiating strategy, were most apparent in the INF negotiations. As discussed above, problems resulted from considering the technical and policy
issues involved in isolation. These difficulties were compounded by the
administration's embrace of a more ambitious standard of verification
before negotiations began. Although never clarifying precisely how "effective" verification would differ from the standard of adequacy deemed
necessary for previous agreements, administration officials left no doubt
that it would be more rigorous. At the same time, they insisted that the
scope of new arms reduction agreements be expanded to areas that were
inherently more difficult to monitor, such as nondeployed missiles.

Considerable weight was placed by administration officials on the role
of highly intrusive on-site inspections to address the apparent mismatch

between ambitious INF negotiating objectives and the inherent limitations of national technical means of intelligence gathering. The particulars of this intrusive approach were summarized in a State Department release on March 12, 1987. When the Soviet Union accepted many of these measures in principle, second thoughts about the likely costs and benefits of on-site inspections prevailed in the Reagan administration's deliberations. Because there was a legitimate reason for a conservative Republican president to scale back his verification proposals (Soviet acceptance of the "global double zero" solution), this move caused President Reagan only modest political embarrassment while providing the Kremlin with the benefits of a role reversal on the verification issue. A less politically conservative administration would have suffered far more serious costs in similar circumstances.

On the other hand, the Reagan administration succeeded in establishing an impressive INF verification regime despite irregularities in the process of formulating U.S. proposals and negotiating them with the Soviet Union. This regime includes carefully circumscribed inspection measures of unprecedented scope that will undoubtedly serve as the baseline for subsequent accords. Nonetheless, the Reagan administration's INF negotiating experience therefore suggests several areas of caution for the future. Technical and policy issues relating to verification are best dealt with in an integrated fashion as early in negotiation as possible. U.S. national security interests are best served when the verifiability of arms control proposals are carefully assessed *before* selecting negotiating objectives and *before* entering negotiations. If an administration's declaratory policy toward verification is disconnected from other central negotiating objectives, severe difficulties can result. What if the units of account chosen in strategic arms reduction negotiations for their military significance do not lend themselves to professed verification standards? After lengthy negotiations, does an administration admit that its hard work was all in vain? Or, given the pressures that build up to sign an agreement over the course of negotiations, does an administration accept verification provisions that are less than adequate? Skeptics of arms control contend that these pressures can become extraordinarily severe over time. If they are correct, verification issues are particularly ill-suited for a negotiating end game.

Executive Branch Organization for SALT Compliance

Shortly after the SALT I accords were signed, President Nixon directed the secretary of defense to establish an internal mechanism to review U.S. military research, development, and testing for conformity with the SALT agreements, while the Central Intelligence Agency devised a review mechanism for Soviet programs. Both agencies restricted sharply the role played by other bureaucracies. There were sound reasons for maintaining OSD and CIA primacy in their respective review groups, but the limits placed on interagency coordination also reflected the basic organizational drive to control the bureaucratic action.

In January 1973, an OSD directive established a focal point for SALT reviews within the office of the director of defense research and engineering. This office was authorized to develop internal procedures for SALT monitoring with the "advice and assistance" of the general counsel's office within the Department of Defense. Classified revisions to the original Defense Department directive were promulgated in 1977, providing further guidance on testing programs based on the Pentagon's internal criteria. Each military department and designated defense agencies were directed to supply quarterly reports on SALT compliance and to designate a SALT monitoring "element" within their ranks. The SALT compliance assessment office, in turn, was directed to review these reports and notify the secretary of defense that U.S. programs were proceeding in compliance with the SALT agreements.[6] Agencies outside the Pentagon have had no formal role in the process of monitoring SALT-related compliance questions.

These internal regulations were applied until December 22, 1986, when Secretary of Defense Caspar Weinberger signed a memorandum rescinding those portions pertaining to the SALT I Interim Agreement and the SALT II treaty.[7] This memorandum implemented President Reagan's announcement of May 27, 1986, that, in response to "the continuing pattern of Soviet noncompliance," the United States would no longer base its strategic force posture decisions on the provisions of these agreements, but on the basis of strategic need and estimates of the Soviet threat.[8]

Internal Pentagon regulations concerning compliance continued to be applied to programs relating to the Anti-Ballistic Missile (ABM) treaty, even after the Reagan administration's decision that a "broad" interpretation of the treaty's terms was warranted. These evaluations are undertaken by the same office within the office of the under secretary of defense for acquisition that used to conduct formal SALT program reviews. Participants include representatives of the DOD general counsel's office, the office of the assistant secretary of defense for international security policy and the office of the Joint Chiefs of Staff to review compliance issues.[9] Even before the Congress mandated annual reports on the conformity

---

[6]Department of Defense Directives 5100.70 and 5100.72, January 9, 1973, "Implementation of Strategic Arms Limitation (SALT) Agreements."

[7]"Adherence to and Compliance with Agreements," U.S. Arms Control and Disarmament Agency, February 11, 1987, 1-2.

[8]Fact Sheet, "U.S. Interim Restraint Policy: Responding to Soviet Arms Control Violations," The White House, Office of the Press Secretary, May 27, 1986.

[9]See "Report to the Congress on the Strategic Defense Initiative," Strategic Defense Initiative Organization, April 1987, Appendix D, D-3–D-4.

of the Strategic Defense Initiative (SDI) programs with the ABM treaty in 1984, the primary focus of the Pentagon's internal ABM treaty compliance evaluations became the test activities of the SDI organization.[10] These reports are drafted and coordinated by the Pentagon's internal compliance assessment office. Unlike its earlier SALT compliance assessments, the annual SDI compliance report is circulated in draft form to other agencies within the executive branch for coordination, although no formal requirement exists for doing so.

Just as the Pentagon moved to establish an internal SALT review mechanism, the Central Intelligence Agency quickly preempted the role of other executive branch agencies in reviewing Soviet compliance practices. Director of Central Intelligence Richard Helms did so by establishing a steering group on monitoring strategic arms limitations, chaired by his deputy director of Central Intelligence. The other members of the SALT steering group were the director of the Defense Intelligence Agency, the director of the Bureau of Intelligence and Research (INR) in the State Department, and the deputy director for intelligence at the CIA. The director of the National Security Agency, initially an observer, was later accorded full membership. A working group was also formed, headed by a senior officer of the CIA.[11]

The SALT steering group's main function was to monitor Soviet compliance with the terms of the Interim Agreement and the ABM treaty. It also had important responsibilities in overseeing the tasking of U.S. collection assets and the dissemination of information collected and evaluated to appropriate officials within the executive branch. At first, these assessments were produced on a quarterly basis; eventually, they were produced every six months.

During the Nixon, Ford, and Carter administrations, intelligence community officials attempted to keep as clear a distinction as possible between treaty monitoring, which was their province, and drawing conclusions as to Soviet compliance. The latter, which necessarily had to draw on the negotiating record and additional diplomatic exchanges, went beyond the realm of intelligence collection and assessment and was thus deemed to be the function of policy officials.

As a consequence, the SALT steering group's reports were narrowly drawn to relate Soviet military activities to specific provisions within the accords. On occasion, these activities raised questions concerning compliance or could lend themselves to multiple interpretations—at least until further evidence was collected. When this was the case, the steering group's reports would provide the most likely explanation for the activities, noting

---

[10]See Section 1102 of the Department of Defense Authorization Act, FY 1985 (Public Law 98-525), October 19, 1984.
[11]See Strategic Arms Limitation Hearings, 1927–28.

differences of view within the intelligence community when they existed. No conclusions were drawn as to compliance or noncompliance.

Disputes over Soviet compliance arose soon after the accords were signed, most dramatically over the construction of what appeared to be new missile silos, an activity expressly prohibited by the Interim Agreement. With continued construction and through exchanges with Soviet officials, it became clear that they were command and control modules, not missile silos.[12] When this and other compliance disputes arose in the Nixon administration, the verification panel was not always convened to discuss the SALT steering group's assessments or to deliberate on how U.S. concerns should be transmitted to the Soviets. A special diplomatic channel, the Standing Consultative Commission (SCC), was established in the SALT I accords specifically to address treaty implementation and compliance questions. Initially, it, too, was bypassed by Kissinger who preferred to confer first with a few of his associates and with the president and then raise U.S. concerns directly with Soviet officials.

Much ill will arose from Kissinger's preference for back-channel discussions and his tactics of placing temporary holds on information dealing with Soviet compliance on the part of some members of the verification panel as well as from U.S. negotiators.[13] A spate of leaks airing these issues and asserting Soviet violations appeared in *Aviation Week and Space Technology* and other publications.[14] These assertions were not supported by the SALT steering group's highly classified analysis, but they had considerable political impact. A number of Nixon administration officials who felt most aggrieved by Kissinger's tactics—such as Chief of Naval Operations Elmo R. Zumwalt, Jr. and Chairman of the Joint Chiefs of Staff Admiral Thomas H. Moorer—subsequently became vocal critics of the SALT process and of Soviet compliance practices.

After this rocky start, the executive branch's handling of Soviet compliance questions improved considerably. From 1975 on, questions were first addressed through the interagency process, which resulted in coordinated instructions to the U.S. commissioner at the SCC. Instructions for the SCC delegation were usually drafted by the NSC staff in consultation with the U.S. commissioner after the verification panel working group had drafted the needed issue papers. This process did not create serious rifts within the executive branch, although occasionally sensitive SALT implementation and compliance issues required consideration by the verification panel itself.

---

[12]"SALT I: Compliance, SALT II: Verification," U.S. Department of State, Bureau of Public Affairs, Selected Documents No. 7, Department of State Publication 8939, General Foreign Policy Series 305, March 1978, 4–5.

[13]Pike Committee Hearings, 1927–61.

[14]*Aviation Week and Space Technology,* October 2, 1974, 75–6; November 11, 1974, 77–9; November 25, 1974, 80–1; February 3, 1975, 73–4; January 5, 1976, 67–9; May 24, 1976, 208–9; and May 31, 1976, 190–1.

The Carter administration utilized the same bureaucratic apparatus for handling SALT compliance disputes as its predecessors. The special coordinating committee, like the verification panel, gave little time to compliance issues. A notable exception occurred in 1979, when evidence came to light suggesting a possible Soviet violation of the Biological Weapons Convention. In contrast, SALT compliance issues were handled in interagency working groups backstopping the SCC and the ongoing SALT negotiations. As before, the SALT steering group continued to prepare monitoring reports within the intelligence community on a semiannual basis.

With the prospect of a second strategic arms limitation agreement, the attention of the Carter administration began to focus on how to present to the Congress and to the public its assessment of Soviet compliance practices. The SALT steering group's reports could not be used for this purpose because they contained highly classified material. These reports also did not present conclusions on Soviet compliance practices, nor did they discuss how these issues had been resolved in the SCC channel. The task of preparing less highly classified, as well as unclassified, reports fell to an interagency SALT task force, with officials from ACDA and the OSD taking the lead. Unclassified reports were ultimately published in 1978 and 1979, concluding that whenever the United States raised compliance issues with the Soviets, "in every case the activity ceased, or subsequent information has clarified the situation and allayed our concern."[15]

This conclusion was strongly disputed by critics of the SALT agreements and of the compliance diplomacy practiced by the Nixon, Ford, and Carter administrations, some of whom were offered key positions within the executive branch during the Reagan administration. Their most articulate spokesman, Richard N. Perle, characterized the SCC in a memorandum transmitted by Secretary of Defense Caspar Weinberger to President Reagan as a "diplomatic carpet under which Soviet violations have been continuously swept, an Orwellian memory hole into which our concerns have been dumped like yesterday's trash."[16]

Executive Branch Organization for Compliance in the Reagan Administration

Despite its name, the Reagan administration's Arms Control Verification Committee occupied itself almost entirely with compliance issues since its inception in 1982. The ACVC was supported by an analysis group and policy group with common core membership that developed

---

[15]"Verification of SALT II Agreement," U.S. Department of State, Bureau of Public Affairs, Special Report No. 56, August 1979, 3.

[16]Walter Pincus, "Weinberger Urges Buildup Over Soviet 'Violations'," *Washington Post*, November 19, 1985, A-1.

documents for consideration by the committee. The analysis group was cochaired by the director of ACDA's Verification and Intelligence Bureau and the chief of the CIA's arms control intelligence staff. Its primary responsibility was to correlate the Kremlin's military activities with its obligations under arms control agreements.

The ACVC analysis group did not reach conclusions as to Soviet compliance or noncompliance. This was the responsibility of the ACVC policy group, cochaired by the Department of State's assistant secretary for politico-military affairs and the Pentagon's assistant secretary for international security policy. Participants in ACVC analysis and policy reviews were drawn from the relevant bureaus within the Department of State, ACDA, CIA, and, as necessary, the Department of Energy. The Pentagon was represented by officials from the Office of the Secretary of Defense, the Joint Chiefs of Staff, and the Defense Intelligence Agency (DIA).

The awkward arrangement of separate committees of the same individuals under different cochairmanships was designed, in part, to maintain a distinction between analyzing Soviet military activities on the one hand and arriving at compliance judgments on the other. The division between the two—never entirely clear—became far less distinct than in prior administrations. To begin with, the line between ACVC analysis and policy was often difficult to define. The representative of the director of central intelligence was also placed in the awkward position of attempting to ensure that available intelligence was not misrepresented in the policymaking process, while trying to maintain whenever possible a common intelligence community position. On several of the Reagan administration's compliance concerns, this was particularly hard to do, as the DIA and CIA positions analyzing Soviet compliance practices diverged in important ways.[17] Intelligence community officials have also asserted themselves in policy discussions when they believed "equity" issues to be involved—when policy judgments might adversely impact the U.S. intelligence community's ability to carry out its monitoring responsibilities.

The ACVC's deliberations became the basis for annual administration reports on Soviet compliance practices. These reports, which were also mandated by Congress, drew from the intelligence community's semiannual analyses carried over from the SALT steering group. When the ACVC policy group's deliberations were unable to resolve interagency differences on compliance judgments, the National Security Council staff, presumably acting after appropriate consultation with and instruction from the president, refereed these disputes. A variety of qualifying phrases

---

[17]Michael R. Gordon, "Defense Department is Rebuffed on Soviet ABM Treaty," *New York Times*, March 5, 1987, A–10; R. Jeffrey Smith, "Administration at Odds over Soviet Cheating," *Science*, vol. 228 (May 10, 1985): 695–6; R. Jeffrey Smith, "Scientists Fault Charges of Soviet Cheating," *Science*, vol. 220 (May 13, 1983): 695–7; and Michael Krepon, "CIA, DIA at Odds Over Soviet Threat," *Bulletin of the Atomic Scientists*, vol. 43 (May 1987): 6–7.

were attached to the resulting lists of citations, reflecting the uneven nature of the evidence to support the administration's charges. Cabinet-level interagency reviews on these and other Soviet compliance questions have been rare during the Reagan years, as in prior administrations.

Well-publicized bureaucratic debates took place within the Reagan administration on how to handle these compliance issues through diplomatic channels. Civilian Pentagon officials initially opposed using the SCC channel for SALT II compliance questions, a position that the Reagan administration embraced for over a year. When new disputes arose, the State Department preferred to handle them outside the SCC, as in the early SALT years. At the same time, the Defense Department favored an accusatory approach while demanding a rollback of Soviet malpractices.[18]

The drafting of instructions to the U.S. delegation at the SCC became a source of controversy, although more in the media than within the administration, where the U.S. commissioner had little bureaucratic support to engage in problem-solving dialogues with his opposite number. The failure of the Soviet Union to pursue corrective actions deemed necessary by the Reagan administration became the basis for the president's decision to end his policy of SALT "interim restraint." At a special session of the SCC, called by the Kremlin to learn the meaning of this administration decision, the U.S. commissioner advised his counterpart that compliance discussions relating to the Interim Agreement and to the SALT II treaty would no longer be entertained by the U.S. delegation.[19]

Used successfully in prior years, the SCC did not resolve compliance problems in an administration deeply divided over whether to reaffirm the objectives and purposes of the agreements under its jurisdiction. This is not surprising, because the SCC's "powers" to resolve compliance concerns successfully are no greater than the willingness of both parties to do so. The responsibility for the decline of this institution rested initially on the Soviet Union for badly bending and—in the case of the Krasnoyarsk radar—clearly breaking SALT provisions that were negotiated in good faith. The Reagan administration must also, however, assume a large burden of responsibility for the demise of the SCC. It, too, bent treaty provisions when deemed necessary and declined to adopt problem-solving

[18]Michael R. Gordon, "Officials Say U.S. Agrees to Discuss Arms Policy Shift," *New York Times,* July 15, 1986, A–1; Lou Cannon and Walter Pincus, "U.S. Stance at SALT Meeting Debated," *Washington Post,* July 17, 1986, A–27; Walter Pincus, "U.S. to End SALT II Discussions with Soviet Union Late This Year," *Washington Post,* July 25, 1986, A–23; Michael R. Gordon, "Administration Divided on Value of Arms Pacts," *New York Times,* July 23, 1986, B–8; and Michael R. Gordon, "An 'Orwellian Memory Hole'," *National Journal,* vol. 18, no. 42 (October 18, 1986): 2527.
[19]Michael Krepon, "How Reagan Is Killing A Quiet Forum For Arms Talks," *Washington Post,* August 31, 1986, D–1.

approaches to address its stated concerns.[20] Given the strong antipathy to SALT held by some Reagan administration officials and the unwilling-ness of others to make the resolution of SALT compliance problems a pri-ority, these agreements eroded steadily over time. It is difficult, therefore, to envision any alternative bureaucratic arrangements that would have yielded a significantly different result.

Recommendations

It is a truism that good people can overcome the constraints of poor bureaucratic structure, while the best organizational design can be frus-trated by poorly motivated or incompetent individuals. These caveats aside, the organization of the executive branch does matter—particularly for verification and compliance issues. Future administrations will find the inherent problems of arms control verification and compliance diffi-cult enough without compounding them with bureaucratic complications. Some of the lessons that may be culled from the experiences of the Nixon, Ford, Carter, and Reagan administrations follow.

Maintain a Centralized, NSC-staff Directed System. The potential for abuse clear-ly exists in an overly powerful NSC adviser and overzealous NSC staff as well as in a decentralized system of "cabinet government." Both were on display in the Reagan administration. Although actions taken by the NSC staff harmed the president and the country in the Iran-Contra affair, in verification and compliance matters the administration wounded itself by adopting decentralized bureaucratic procedures. Without strong NSC staff direction, bureaucratic disputes were not resolved, impasses were a natural occurrence, and technical analysis and policy judgments were not well integrated. Enduring disputes between cabinet officers also meant that important and politically sensitive verification issues remained un-decided until the final phases of the negotiations.

Although acknowledging that presidents will organize the executive branch in ways that best suit their individual needs, the Tower Commis-sion wisely concluded that the interagency process "generally operates better when the committees are chaired by the individual with the greatest stake in making the NSC system work."[21] Contentious verification and com-pliance issues have a better chance of being resolved in a timely, well-integrated fashion when interagency committees are chaired by the na-tional security adviser or his chosen deputy. In this area, the benefits of

---

[20]Leo Sartori, "Will SALT II Survive?" *International Security*, vol. 10, no. 3 (Winter 1985/86): 147–74; James A. Schear, "Arms Control Treaty Compliance: Buildup to a Breakdown," *Inter-national Security*, vol. 10, no. 2 (Fall 1985): 141–82; and Michael Krepon, "Look Who's Snubbing SALT Now," *Washington Post*, December 16, 1986, A-19.

[21]"Report of the President's Special Review Board," V-5.

a centralized NSC staff-directed system appear to outweigh the attendant risks.

Maintain Annual Interagency Reviews on Verification Resource Allocation Questions. Intelligence analysts on the Soviet Union, arms control negotiators, technical experts responsible for devising new intelligence collection devices, budget officials, and high-ranking policymakers have responsibility for different aspects of the same problem: how to monitor effectively Soviet military developments that can affect the security of the United States. Yet policymakers know relatively little about the process and difficulties of intelligence collection, while intelligence analysts have few opportunities to discuss resource requirements with those who make policy. Resource allocation considerations are a constant concern for some intelligence community officials, but remain a periodic concern for policymakers. There are many claims on intelligence collection systems as well as analysts; no collection system exists solely for the purpose of monitoring arms control agreements. Thus, difficult choices must constantly be made between claimants.

Interagency meetings on resource allocation and verification policy issues need to occur on a regular basis to correlate U.S. negotiating and policy objectives with projections of Soviet military efforts and with investment strategies for U.S. collection assets. They can be most useful if policymakers and intelligence community officials with diverse but interrelated functions could attend. A semiannual interagency meeting could provide a forum that helps integrate U.S. negotiating and verification strategies while providing useful insights for monitoring investment decisions in the future. Because the CIA director has the overall responsibility to implement decisions taken, he would be the appropriate person to chair these meetings.

Increase the Role of ACDA's Verification Bureau in Research; Decrease it in Verification Policy. The periodic establishment and disestablishment of a separate ACDA verification bureau (one of four within the agency) has always been a source of political controversy. A separate bureau makes sense, but not if it competes with or duplicates the work of ACDA's functional bureaus. Instead, an appropriate model might be the creation of the SALT support staff—a modestly-sized group within the intelligence community that drew on personnel with functional responsibilities elsewhere in the agency. Analogously, verification considerations might be best integrated into decisionmaking when the strategic affairs, multilateral affairs, and nonproliferation bureaus within ACDA continue to have this responsibility.

Because of its small size, ACDA is one federal agency that can be responsive to redirection from the top. Given the political passions generated by arms control efforts, the agency is subject to dramatic mood swings with changes in administrations. Thus, the more deeply ACDA delves into the analysis of Soviet compliance, the more politicized those assessments will become, whether the stimulus comes from supporters or critics of arms control. Those who have been pleased by the results of ACDA's cochairmanship of the ACVC analysis group during the Reagan administration might consider the consequences of a similar role under a liberal democratic administration. And those who generally wish ACDA to take a higher profile in interagency deliberations must ask what institutional benefits derive from taking a lead role in compliance assessments. On balance, ACDA's role in this area should be drastically curtailed.

What, then, should ACDA's verification bureau do? This bureau's personnel would be much better utilized in identifying problems that require additional study and in helping to direct resources toward their solution. Some verification problems cut across functional areas, while the fruits of negotiation in one area might have some utility elsewhere. ACDA is an ideal place for these cross-connections to be identified and developed further. It would also be useful to have ACDA direct its attention more to problems in verification that go beyond the time-lines most policymakers are able to address.

The allocation and distribution of research funds for verification within the executive branch require review. The Department of Energy annually spends approximately $100 million for verification research. Most of these funds have been expended for verification of nuclear weapons test limitations at the three national laboratories primarily engaged in the design of new nuclear weapons: Los Alamos, Lawrence Livermore, and Sandia. In contrast, ACDA's entire total research budget is approximately $1 million, one quarter of which is directed toward verification issues. Clearly, verification research activity within the executive branch is unbalanced with respect to disbursing agencies, research topics, and recipients. Any reallocation of resources within the executive branch applied to verification should strengthen ACDA's role and aim toward the development of additional centers of verification expertise upon which the nation can draw.

Return the Compliance Analysis Function to the Intelligence Community. The analysis of Soviet practices relating to compliance, which was formerly coordinated by the Central Intelligence Agency, was codirected by the CIA and ACDA during the Reagan administration. At the same time, the intelligence community necessarily became more centrally involved in compliance policy decisions as a result of interagency disputes over Soviet noncompliance. The Reagan administration's noncompliance reviews were improved by CIA participation, which counterbalanced some of the more politicized assessments offered by OSD and ACDA. Still, the compromise conclusions selected by the Reagan administration to bridge interagency differences did not always accurately reflect the available intelligence. The clearest case of this involved the administration's repeated assertions of "likely" Soviet violations of the Threshold Test Ban Treaty.[22]

The analysis of Soviet military practices, whether associated with arms control agreements or not, is the proper function of the intelligence community. This function is best carried out by the intelligence community, not by an interagency process cochaired by ACDA and the CIA. Judgments of Soviet compliance, which require evaluation of the negotiating record and subsequent diplomatic exchanges, are the appropriate province of policymakers, assisted by counsel, diplomats, and intelligence community staff, as needed. These compliance judgments, like other policy issues, are best addressed by interagency reviews, chaired by the NSC staff. If disputes persist in this process, they must be elevated to the president for review and decision. The intelligence community has several roles during this process. One is to make sure that supporting intelligence is as objective as possible, not misrepresented, and fully coordinated. When differences exist within the intelligence community, they must be highlighted and the reasons for those differences explained. Policymakers also need to know the degree to which intelligence community assessments are sensitive to new data.

Create Bureaucratic Checks against the Politicization of Compliance Reports. Bureaucratic checks against overdramatizing or downplaying compliance problems are obviously desirable, but extremely difficult to devise. When compliance issues are downplayed by administration officials, they can expect to be excoriated by some members of Congress and media outlets. In contrast, government officials who provide overdrawn assessments of Soviet noncompliance face fewer gauntlets because it is easier for them to assert widespread Soviet cheating than for others to disprove it. Either way, the Congress's ability to promote higher standards of reporting will likely remain modest. Periodically, members of Congress, particularly those on the select committees on intelligence, have asked administration officials about their compliance judgments, intelligence community officials about their analysis, and why the two may vary. In practice, however, the Congress's prerogative to call administration witnesses and to ask potentially embarrassing questions can serve this function only to a limited extent, particularly because the record developed in closed session remains classified.[23]

---

[22]Committee on Foreign Relations, Ex. N, 94–2, "Threshold Test Ban Treaty and Peaceful Nuclear Explosions Treaty," Hearings (Washington, D.C.: GPO, 1987); R. Jeffrey Smith, "Dispute over Soviet Testing Heats Up," *Science*, vol. 228 (May 31, 1985): 1072.

[23]A notable exception to this general rule is the work of Congressman Dave McCurdy. See "Intelligence Report to Arms Control, Report by the Permanent Select Committee on Intelligence of the House of Representatives," together with dissenting views, (Washington, D.C.: GPO, November 19, 1987).

Within the executive branch, two institutions already exist that could conceivably provide greater protection against overdramatized or weak reporting on Soviet compliance practices. Within the executive office of the president, the President's Foreign Intelligence Advisory Board (PFIAB) was created to "review the performance of all agencies of the Government that are engaged in the collection, evaluation, or production of intelligence or the execution of intelligence policy."[24] Because the quality of an administration's reports on compliance bears on the work of the intelligence community and the integrity of its assessments, the PFIAB could conceivably review this process, if it wishes to do so.

The General Advisory Committee (GAC), ACDA's board of advisers, also has a broad enough mandate to review compliance assessments. Indeed, the GAC made such a review the focus of its activities during President Reagan's first term in office. The quality of the GAC's analysis was questioned by many experts, however, as the GAC found numerous "material breaches"—including "breaches of authoritative unilateral commitments, whether written or oral"—that the best efforts of previous administrations had failed to detect.[25] The composition of the GAC, which included some individuals with little background in these issues and others with firmly held views concerning the expansive nature of Soviet noncompliance, further undermined the credibility of its analysis.

An advisory body that wishes to investigate the compliance issue in depth should include individuals with distinguished backgrounds in intelligence, military affairs, and diplomacy. Retired members of Congress might also usefully serve on such a panel.[26] All should have some familiarity with the esoteric issues involved, with few axes to grind. If an advisory body wishes to conduct inquiries over the handling of compliance controversies in the past or to investigate current disputes, it should have the power to call expert witnesses within the intelligence community. If the composition of an advisory body is exceptional enough, and if it has direct access to the president, its reports can help provide needed perspective on the compliance issue and lend credibility to executive branch reports. To maintain a nonpartisan status, membership of such an advisory body should be staggered to overlap changes of administrations.

Improve Annual Executive Branch Reports on Soviet Compliance Practices. Unfortunately, the compliance issue has become overly politicized. The best corrective at this point is more balanced information, not benign neglect.

---

[24]Executive Order 12331—President's Foreign Intelligence Advisory Board, October 20, 1981.

[25]"A Quarter Century of Soviet Compliance Practice under Arms Control Commitments: 1958–1983," General Advisory Committee on Arms Control and Disarmament (Washington, D.C.: The White House, October 1984), 3.

[26]The author is grateful to Robert Einhorn for this idea.

An advisory body such as the one already described can play an important function in this regard, as could the annual reporting requirement
established by the Congress. Reports in the future, unlike those issued
by the Reagan administration, might best serve the public interest by placing compliance issues in a broader context and by discussing the military
significance associated with each of the Kremlin's questionable practices.

The Reagan administration's annual reports only provide citations
of noncompliance, and they barely differentiate between these citations
as to their potential military significance. A broader context of executive
branch reporting would include the transmittal of information on when
the Kremlin's behavior has been satisfactory or exemplary, as well as when
it has not. The executive branch's semiannual reports assessing compliance with the Helsinki Final Act adopt this approach. The purpose of
broadening the scope of compliance reports is not to provide a balance
sheet or to excuse Soviet misbehavior, but to allow the public and the
Congress to assess for themselves the overall picture of Soviet compliance
practices. In this regard, annual executive branch reports might best serve
the public interest if they included some evaluation of the military significance attached to specific instances of Soviet noncompliance that are
detected. Again, the purpose would not be to condone Soviet noncompliance that has little or no military significance, but to help readers to
consider what responses would be most appropriate.

*Provide Interagency Coordination of Evaluations of both U.S. and Soviet Compli-
ance Practices.* At present, policy judgments within the executive branch
on Soviet compliance practices are coordinated among the appropriate
bureaucracies, while U.S. government assessments of its own compliance
practices remain largely the domain of the Department of Defense. Related
issues that may arise, such as the case of the Reagan administration's reinterpretation of the ABM treaty, are often handled on an ad hoc basis. In
this case, the reinterpretation was initiated by the Pentagon, pursued by
the State Department's legal adviser, and addressed irregularly in interagency channels, resulting in a decision before a thorough review of the
negotiating record and subsequent U.S.-Soviet exchanges on the subject
had been conducted.[27] Regularized interagency procedures might not have
prevented this occurrence, but they can help minimize confusion or embarrassment, if administration officials wish to do so.

The absence of common procedures for assessing U.S. and Soviet compliance practices can also lead to a double standard in judging superpower
conduct, something the Reagan administration has repeatedly warned
against. Is the SS-25 a brand new missile while Thule and Fylingdales are

---

[27]Raymond L. Garthoff, *Policy Versus the Law: The Reinterpretation of the ABM Treaty* (Washington,
D.C.: The Brookings Institution, 1987).

not brand new radars? Does the unilateral Soviet interpretation of the new missile "types" rule for SALT II constitute an impermissible violation, while the unilateral Reagan administration reinterpretation of the ABM treaty is permissible?

If administration officials wish to avoid a double standard on these issues, they would be wise to establish an interagency review process for U.S. programs comparable to the one that currently exists for evaluating Soviet compliance. Officials from the office of the general counsel at the Department of State, Defense, and ACDA should participate in all such interagency reviews. Reviews of U.S. programs should continue to be chaired by the office of the under secretary for defense research and acquisition, acting as executive agent for the NSC. This office is best equipped to lead internal reviews, which also provide an opportunity for oversight into the SDI program.

6

RULES OF THE ROAD FOR SPACE OPERATIONS

BY PAUL B. STARES

The growing military use of space represents one of the most potent sources of superpower competition and confrontation. At the forefront is the current disagreement about the freedom to develop, test, and deploy strategic defense systems in space. Less prominent, but still important, is the related issue of antisatellite (ASAT) weapons development. As satellites have become more useful for the support of military missions, so the incentives to deny their benefits to an adversary in wartime have grown. So too has the need to safeguard these increasingly valuable assets from attack.

Ensuring the safety of satellites while preventing the hostile use of space, however, poses a major dilemma to U.S. defense planners. The deployment of ASAT weapons will almost certainly encourage the Soviet Union to do likewise, which in turn could undermine the security of U.S. satellites. Although space systems can be fitted with a variety of defensive countermeasures to help protect them in wartime, it is unclear how long these measures can remain effective against a determined and uninhibited adversary. Because the United States is generally considered to have a higher overall dependency on space systems in wartime than the Soviet Union, it is also unclear whether attacks on U.S. satellites could be deterred by the threat of reciprocal action. As a result, many question whether it is in the interest of the United States to risk becoming involved in an ASAT competition with the Soviet Union.

A more ominous consequence of ASAT development could be its effect on crisis stability. The vulnerability of early warning and communications satellites, which both superpowers use to detect a missile attack and make possible effective retaliation, could add dangerous uncertainties to their calculations in a severe crisis. One side may fear that a preemptive ASAT attack coordinated with other military actions could leave it at a decisive disadvantage at the outset of war. Such fears would increase the pressure to strike first if war were considered inevitable or even likely. In view of these concerns, many believe that there are equally strong incentives to constrain the development of antisatellite weapons.

Attempts to limit ASAT systems, however, have not been successful to date. Some progress was made during bilateral negotiations between the United States and the Soviet Union in 1978–79, but they were eventu-ally suspended without result.[1] While the Soviet Union has called persis-tently for a comprehensive prohibition on space weaponry since 1981, and also observed a unilateral ASAT testing moratorium since 1983, the Reagan administration has resisted any attempts to resume negotiations on this subject. The U.S. Congress, however, has restricted the testing of the U.S. antisatellite weapons system now in the advanced stages of development.

The officially stated reasons for the U.S. opposition to space arms control are that ASAT limits would be undesirable and infeasible. It is ar-gued that the United States needs an ASAT capability to deter attacks on its satellites by the threat of retaliation in kind and also to counter the threat posed by Soviet space systems—notably those used to detect, track, and target U.S. naval shipping—in wartime. At the same time it is argued that placing limits on weapons that have been specifically designed to at-tack satellites ("dedicated" ASATs) would achieve very little because there is a whole class of other weapons that have "residual" antisatellite capa-bilities. These include intercontinental ballistic missiles, antiballistic mis-sile interceptors, and electronic jammers, all of which would remain outside of an ASAT agreement. Thus constraints on dedicated ASAT weapons would still leave satellites exposed to the threat from these other weapon systems. Verifying compliance with an ASAT agreement also would pose immense problems, not least because of the similarities between potential weapons-related activities and other commonplace operations in space.[2]

Perhaps the most significant, though unstated, obstacle to progress is Washington's belief that an ASAT arms control agreement would limit its freedom of action to test and develop ballistic missile defense technol-ogies in space as part of the Strategic Defense Initiative (SDI) program. Because the techniques under study for intercepting satellites and ballis-tic missiles in space are similar, constraining ASAT development without hindering the SDI would be difficult. Such restrictions are clearly unac-ceptable while the United States remains committed to developing stra-tegic defenses.

Because the prospects for negotiated limits on antisatellite weapons are so gloomy, many have been attracted to the more modest goal of es-tablishing "rules of the road" for space operations in order to reduce the

---

[1]For discussion of the ASAT negotiations between the United States and Soviet Union during the Carter administration see Paul B. Stares, *The Militarization of Space: U.S. Policy 1945–1982* (Ithaca, New York: Cornell University Press, 1985), 192–9.

[2]These arguments are laid out more fully in Ronald Reagan, "Report to the Congress on U.S. Policy on ASAT Arms Control" (March 31, 1984). For a detailed critique of the Reagan administration's arguments see Paul B. Stares, *Space and National Security* (Washington, D.C.: The Brookings Institution, 1987).

negative consequences of space weapon deployments. Rules of the road can be defined as internationally approved regulations or procedures for the operation of spacecraft. Such agreed formal regulations may be useful for four reasons.

First, rules of the road could help reduce the likelihood of misunderstandings and, ultimately, conflicts arising as a result of ambiguous incidents in space. They would achieve this by lowering the probability of potentially dangerous events occurring and also by providing clear and prompt channels of communication for exchanging information if such events do occur.

Second, rules of the road could help reduce the fear of surprise attacks in space by making the intentions and actions of nations more transparent. For instance, they might improve the ability of each side to detect, track, and identify objects in space.

Third, rules of the road could add to the effectiveness of unilateral measures taken to improve the survivability of satellites by inhibiting the freedom of action to attack space systems.

Fourth, rules of the road could provide the basis for additional arms control agreements and set useful precedents for cooperation in other uses of space whether scientific, commercial, or military in nature. They might also prove useful if constraints were placed on the development of space weapons at some future date.

Potential Rules of the Road

The primary purpose of this chapter is to assess a variety of measures that would be considered candidate rules of the road for space operations. The first section examines each of the candidate measures for its merits and possible liabilities; the second assesses whether sufficient technical aids are available to make rules of the road practicable. The third section looks at the potential spin-offs from agreements of this sort. The chapter concludes with an overall evaluation of the desirability and benefits of establishing rules of the road in space.

A wide variety of measures can be considered rules of the road. Although discussed separately, they should not be viewed as mutually exclusive. The effectiveness of some measures may even be enhanced by being combined with others.

A Prohibition on the Use of Force in Space

One of the most basic rules of the road could be an agreement that bans the use of force in space as well as hostile interference with the functioning of spacecraft. Although this agreement would do little to protect satellites in time of war, it could be beneficial in peacetime and during a severe superpower crisis. Such a prohibition, for instance, might deter brinkmanship and other acts of harassment in space by providing a distinct legal threshold that states would have unambiguously to cross.

To some extent interference with satellites and the use of force in space are already prohibited by international agreement. The Outer Space Treaty signed in 1967 explicitly states that space activities be carried out in accordance with international law, including the United Nations Charter, thereby indirectly prohibiting the use of force in space except for reasons of self-defense.[3] A more specific, though still ambiguous, limitation on harmful interference with satellites is contained in the 1972 SALT I agreements. Both the Interim Agreement on the Limitation of Strategic Offensive Forces and the Anti-Ballistic Missile treaty include articles in which the United States and the Soviet Union agree "not to interfere with the national technical means of verification of the other Party." Although a commonly accepted definition of national technical means has never been reached, it is generally understood to include reconnaissance satellites.[4] Furthermore, the 1971 and 1984 Direct Communications Link or "hot line" improvement agreements include the provision that both parties "agree to take all possible measures to assure the continuous and reliable operation" of the network's satellite communication links and terminals. The 1973 International Telecommunication Convention also obligates each party "not to cause harmful interference to the radio services or communications of other Members. . . ."

Despite the considerable body of existing international law that places constraints on the use of force in space, much of it is ambiguous or narrowly defined. Negotiating a more explicit and inclusive prohibition could encounter some significant problems, however. Defining at the outset what constitutes such terms as the "use of force" or "hostile interference with space objects" could be difficult. For example, should a prohibition refer only to those acts that damaged or destroyed spacecraft, or should it also include those that impeded the communication links to and from satellites? Should allowances be made for "legitimate" uses of force in space, such as acts of self-defense in which parties to the agreement believed that their national security was threatened by other activities in space? Such rights of circumvention were reportedly demanded by the Soviets during the 1978–79 bilateral ASAT negotiations. There is the additional question of whether such an agreement, and for that matter other rules of the road, would be confined strictly to U.S. and Soviet space operations or whether they should also include the spacecraft of other nations. Again, the Soviets reportedly pushed for an exclusive accord in this regard, raising in the process concerns about the status of allied space systems.

---

[3]For the full texts of the agreements cited here see *Space Law: Selected Basic Documents*, 2nd ed., Senate Committee on Commerce, Science and Transportation, 95th Congress, 2nd session (Washington, D.C.: GPO, 1978); and *U.S. Arms Control and Disarmament Agreements: Texts and Histories of Negotiations* (Washington, D.C.: Arms Control and Disarmament Agency, 1982).
[4]For a discussion of the possible ambiguities surrounding this clause see Stares, *The Militarization of Space*, 165–6.

Although these questions would need to be resolved, they do not represent insurmountable problems. A tentative outline for a prohibition on the use of force in space was apparently accepted prior to the breakdown of the bilateral negotiations in 1979.[5] The Soviets have also included such a provision in the various draft treaty proposals banning space weapons that they have submitted to the United Nations General Assembly. The question remains, however, whether they would consider a separate prohibition on the use of force to be implicitly condoning and legitimizing the deployment of space weaponry.

An "Incidents in Space" Agreement
The principal function of an incidents in space agreement is to reduce the likelihood of accidents and other incidents occurring in space that might cause a serious deterioration in superpower relations and even lead to hostilities. Similar agreements are already in place for air operations (the 1944 Chicago Convention) and for maritime activities (the 1960 International Regulations for Preventing Collisions at Sea). The most obvious model for an incidents in space agreement, however, is the 1972 U.S.-Soviet Agreement on the Prevention of Incidents On and Over the High Seas, which significantly curbed (though it did not eliminate) what had become a growing number of dangerous incidents between the warships of the two superpowers.[6] Although maritime operations differ considerably from space activities, an analogous incidents in space agreement could serve the same function.

One of the principal objectives of such an agreement would be to reduce the risk of collision in space. As the level of space activity increases, so this problem will inevitably grow. Although the main threat is from "space junk"—inactive satellites, booster fragments, and other debris from space operations—which is difficult to detect and therefore avoid, other, more serious sources of collision could develop in the future. Close inspections of space objects to determine their function and characteristics could become quite common if either superpower deploys weapon systems in space. The risk of inadvertent collision would almost certainly grow as a consequence. Similarly, if weapon systems begin to be regularly tested and deployed in space, the chances of accidental damage to other spacecraft will also increase.

More worrisome is that both superpowers may begin to view space as an arena where brinkmanship can be conducted, leading to acts of harassment and interference with each other's space activities. Worse still would be simulated attacks against spacecraft in order to discover what

---

[5]Ibid., 200.
[6]Sean M. Lynn-Jones, "A Quiet Success for Arms Control: Preventing Incidents at Sea," *International Security*, vol. 9, no. 4 (Spring 1985): 154–84.

protective measures had been adopted to improve survivability. For example, a high-speed "fly-by" in close proximity to another satellite or weapons platform might uncover that it had an emergency maneuvering system or possessed decoys to deceive ASAT sensors. Intelligence of this kind would be very useful for wartime planning. In periods of high tension such maneuvers might be construed as a prelude to an attack, which could cause the crisis to escalate in unforeseen and undesirable ways.

Several measures could be adopted by the superpowers to reduce the risk of incidents in space either causing or escalating a crisis. To reduce the probability of collision, radio beacons would be attached to active satellites to aid their detection by ground-based sensors. These would essentially serve the same purpose as navigation running lights on ships. Minimum separation distances between spacecraft could also be stipulated as well as advance notification of space activities.

An incidents in space agreement could also include a prohibition on dangerous maneuvers in the proximity of the other side's spacecraft as well as simulated attacks. This prohibition would encompass the launching of objects into the immediate path of passing satellites to cause a reaction or otherwise impede the functioning of the spacecraft.[7] Deliberate acts of harassment such as close fly-bys, electronic jamming, and interference with a satellite's optical sensors by lasers would also be banned.

As with the incidents at sea agreement, a consultative forum could be established to discuss incidents in space. Alternatively, the Standing Consultative Commission (SCC) could serve the same purpose. Although this forum would discuss space issues ex post facto, more immediate channels of communication to exchange information promptly after an incident or, better still, before it had occurred, also could be established.

Some of these suggested components of an incidents in space agreement might be rejected by the United States or the Soviet Union. The idea of placing radio beacons on spacecraft, for instance, would probably be criticized for aiding enemy targeting in wartime and for foreclosing the deployment of deceptively based spares, sometimes referred to as "dark satellites," that could be activated in an emergency. Because radio beacons also would not address the most likely source of collisions, namely the proliferation of space junk, they are unlikely to be adopted. Advanced notification of military space activities is also likely to be opposed on the grounds that it might allow enemy intelligence the opportunity to monitor events that might not have been possible without forewarning. A prohibition on directing lasers at each side's spacecraft might also compromise the current practice of illuminating satellites with low-intensity lasers to aid

[7]See Donald L. Hafner, "Approaches to the Control of Antisatellite Weapons," in William J. Durch, ed., *National Interests and the Military Use of Space* (Cambridge, Massachusetts: Ballinger Publishing Co., 1984), 264.

photographic identification. A specific limit on the brightness of lasers (a function of their power, wavelength, and the diameter of the primary beam directing mirror) could conceivably avoid this problem, however.

Even if these objections were sustained, there is still merit to an agreement that prohibited dangerous maneuvers and deliberate acts of harassment in space and also established dedicated communication channels to discuss each side's space activities.

"Keep-Out" Self-Defense Zones

More specific regulations governing minimum separation distances between satellites, often referred to as either "keep-out zones" or "self-defense zones," represent another potential type of rules of the road agreement. Delineating areas of space around satellites that an aggressor would need to transgress would not only provide unambiguous warning of attack but might also lengthen the time that an aggressor would need to carry out an attack. The early warning and added time could enhance the ability of satellites to defend themselves. Such zones, therefore, are not intended either to prevent ASAT attacks or to substitute for unilateral measures to improve the survivability of satellites.

Most proposals for self-defense zones are directed at protecting satellites in medium to high earth orbits (above 10,000 kms altitude) and most commonly in the geosynchronous (approximately 36,000 kms) and super-synchronous orbits (above 40,000 kms). Self-defense zones in lower altitude orbits are generally considered to be impractical. Rocket propelled antisatellite devices can be employed in a matter of minutes against low orbiting satellites and directed energy weapons, such as lasers, in only a few seconds. Self-defense zones, therefore, would offer little in the way of additional warning time for these satellites to take evasive action.

In contrast, the threat to satellites in higher altitude orbits from ground-based ASAT systems is currently considered to be significantly lower. Kinetic weapons propelled by conventional rocket engines would take between three to six hours to reach the geosynchronous orbit, which in itself provides some warning to threatened satellites. Ground-based lasers would require enormous sources of power and special adaptive optics to attack satellites in the geosynchronous orbit. Such weapons are not considered feasible until at least the next century. It is, therefore, against space-based threats, particularly those designed to be deployed in close proximity to space targets, that self-defense zones are primarily directed. These include, in the near term, nuclear- or conventionally-armed "space mines," and guided kinetic weapons launched from space platforms. Space-based directed energy weapons, such as lasers and neutral particle beam generators, are a long-term possibility.

High altitude self-defense zones have other advantages. The satellites used for detecting a ballistic missile attack and for strategic command and

105

control—in other words, the space systems that buttress nuclear deterrence—are primarily located in these higher orbits and therefore would be the principal beneficiaries of this kind of arrangement. Paradoxically, self-defense zones have also been discussed in the context of safeguarding space-based strategic defense systems.[8]

The most sophisticated proposal for self-defense zones has been put forward by Albert Wohlstetter and Brian Chow.[9] They propose that the geosynchronous orbit be divided into 36 zones representing 10 degrees of arc amounting to 7,400 kms in width. Twelve of these zones would be apportioned each to the Warsaw Pact, NATO, and neutral countries. Similar zones are also suggested for other orbits. Additional provisions to allow a limited number of satellites from each bloc to be stationed in the zones of others, as well as the transit of satellites through zones, could also be agreed upon. Satellites found violating their permitted boundaries would be subject to inspection or attack.

In an environment where ASAT development is completely unconstrained, self-defense zones could help bring some stability to a potentially dangerous situation. They would certainly reduce the scope for a sudden disabling attack against key early warning and strategic communication satellites in the geosynchronous orbit. On further examination, however, unilateral measures may prove just as effective. Furthermore, it is unclear whether the effort to negotiate and then maintain a self-defense zone agreement would really be worth the benefits.

In the first case, it may be simpler and cheaper in the long term to reduce drastically the level of dependence on space systems, either by proliferating the number of satellites that carry out nuclear command and control functions or relying more on ground- and air-based alternatives. The former would depend on how efficient ASAT capabilities become, while the latter would depend on the relative performance of alternative systems. In an unconstrained ASAT environment, self-defense zones may ultimately buy very little time for satellite defensive measures, particularly against sophisticated directed-energy weapons. Some analysts are pessimistic about how much protection even large defense zones might afford satellites.[10]

---

[8]Lieutenant General James Abrahamson, the director of the SDI Office has expressed an interest in "keep-out" zones. See *Department of Defense Appropriations for 1986*: House Committee on Appropriations, 99th Congress, 1st session. (Washington, D.C.: GPO, 1985), pt. 7, 655.

[9]Albert Wohlstetter and Brian Chow, "Arms Control That Could Work," *Wall Street Journal* (July 17, 1985). For a fuller exposition of this proposal see Albert Wohlstetter and Brian A. Chow, *Self-Defense Zones in Space*, Report MDA 903–84–C–0325 (Marina del Rey, California: Pan Heuristics, 1986). For other suggestions see Office of Technology Assessment (OTA), *Anti-Satellite Weapons, Countermeasures, and Arms Control* (Washington, D.C.: GPO, 1985), 136–8.

[10]Leonard Anthony Wojcik, "Separation Requirements for Protection of High Altitude Satellites from Co-Orbital Antisatellite Weapons" (Ph.D. dissertation, Carnegie-Mellon University, 1985), 121–3; OTA, *Anti-Satellite Weapons*, 138.

The negotiation of self-defense zones in space may also run up against some significant legal problems. In particular, an agreement of this kind would in effect extend sovereign rights into space, a development that hitherto has been rejected by international law. Article II of the Outer Space Treaty explicitly states that "outer space . . . is not subject to national appropriation by claim of sovereignty, by means of use or occupation, or by any other means." If an exception were made for self-defense zones, it might nevertheless lend support to the claim of some equatorial nations to sovereign rights of portions of the geosynchronous orbit.[11] It might also reopen the debate about the demarcation between air space and outer space, with implications for the legitimacy of reconnaissance satellite operations as well as the use of low-flying manned spacecraft, such as the shuttle and, in the future, hypersonic aerospace planes.[12] The United States would have to be careful that self-defense zones would not compromise current intelligence gathering and arms control treaty monitoring from the geosynchronous orbit. In short, self-defense zones may provide only negligible benefits against technically advanced threats, with the added possibility of unwelcome side effects.

Crisis Management Procedures

Another measure designed to prevent conflicts arising out of incidents in space would be to institute bilateral emergency consultative procedures. As noted earlier, this could be made part of a general incidents in space package or it could be established separately. Like the 1971 U.S.-Soviet Accident Measures Agreement, such an agreement could include a provision to immediately notify all signatories of accidents, unauthorized activities, or any other unexplained incidents in space that could risk an outbreak of hostilities. Advance warning of potential collisions in space would also be useful. Special communication links could be set up, perhaps between the space tracking and satellite control centers of the two superpowers or, alternatively, use could be made of the recently established nuclear risk reduction centers. Direct links between the space tracking centers may be preferable given the potential for delays with other channels.

It is important to recognize that crisis control mechanisms can have paradoxical side effects. The likelihood of misperceptions over incidents in space may actually be heightened if one state has the right and the dedicated communication channels to request information from another. If a state cannot supply information on an incident in space, either because it genuinely does not know the answer or because to provide one would

---

[11]OTA, *Unispace '82: A Context for International Cooperation and Competition* (Washington, D.C.: GPO, 1983), 43-4.
[12]There is currently no internationally accepted legal definition of what constitutes outer space.

compromise information about the capabilities of its own satellites and monitoring facilities, the agreement may increase rather than reduce the room for suspicions.[13] Although the possibility of such unintended side effects cannot be dismissed, trust and confidence in the system would hopefully grow over an extended period, especially if a prospective accident were successfully avoided or an ambiguous incident in space were explained satisfactorily.

Cooperative Monitoring Procedures and Confidence-building Measures
    One measure that could help in the monitoring of space activities and go some way in reducing suspicions of hostile intentions in space would be an agreement stipulating prior notification of all space launches. Under the terms of the 1976 UN Convention on the Registration of Objects Launched into Outer Space, satellite launching states are currently obliged to provide the UN registry with the date, territory, and location of the launch, the basic orbital parameters of the space object, an appropriate designator, and information on the general function of the satellite, but only *after* the launch has occurred. This agreement could be modified (as permitted under Article IX) to include advance notification, or a separate bilateral arrangement could be reached between the United States and the Soviet Union for confidential prior notifications. The exchange of more detailed information on the nature and specific function of the spacecraft to be launched may also be feasible. Furthermore, a complete catalog of space objects above a specific size could be jointly generated and maintained either bilaterally between the superpowers or among the existing space nations.

    Although a pre-launch notification agreement would not relieve the national monitoring agencies of the necessity for constant vigilance, it could make their task somewhat easier and perhaps even cheaper. Also, like an agreement prohibiting the use of ASAT weapons, it would in effect provide another legal barrier for states to transgress unambiguously if they harbored hostile intentions in space. The common postponements associated with launch activities could provide a source of irritation and suspicion, but this could be avoided by stipulating that states must promptly notify others of any postponement and provide updated schedules for the next launch.

    Another method to help reduce suspicions of the purpose of orbiting objects is to allow some form of in-space inspection. There is already some precedent for this. The Outer Space Treaty includes provision for the inspection of installations, facilities, and equipment on the moon and other celestial bodies by representatives of other states party to the agreement

---

[13]For a general discussion on this problem see Thomas C. Schelling, "Confidence in Crisis," *International Security*, vol. 8, no. 4 (Spring 1984): 55–66.

on a basis of reciprocity and upon "reasonable advance notice."[14] Physical inspection of space objects would almost certainly be rejected by all nations, but external inspections from a predetermined distance and with prior notification might be acceptable.[15]

The attractiveness of this kind of proposal would obviously depend on whether current capabilities for determining the functions of spacecraft could be enhanced significantly under a cooperative regime. For example, monitoring might be improved by close proximity inspections and using certain surveillance techniques such as x-ray devices that might otherwise be considered provocative. Although this might help deter deliberate violations of an arms control agreement that banned specific activities, procedures for in-space inspection, similar to some of the crisis control measures already noted, could become a potential source of friction. The use of such devices would almost certainly fuel suspicions that close inspections in space would be used for intelligence "fishing expeditions." Third-party involvement in the inspection process might alleviate this concern, but it is doubtful whether this would be acceptable to either superpower. Some might also fear than an antisatellite attack might be carried out under the pretext of inspection. As this is realistically a one-time-only possibility, the risk does not appear to be great, especially if parties avoid becoming critically dependent on a single space system.

Negotiating a space inspection regime could run into some significant obstacles. Determining the amount of prior notification is a tricky issue; too much could provide enough time to conceal prohibited activities while too little could raise suspicions in a crisis. The discussion of permitted inspection techniques could also risk compromising sensitive information about each side's satellites. On balance, unless the verification of a space arms control agreement promised to be significantly enhanced by the right to conduct inspection in space, it is questionable whether it is worthwhile pursuing separately.

A related confidence-building measure would be an agreement to notify all states operating satellite systems of an impending space weapons test. The purpose would again be to avoid accidents in space and reduce the likelihood of misperceptions arising from incidents in space. Again, there are precedents for this type of agreement. The Accident Measures Agreement calls on the two parties to inform each other of planned missile launches that extend beyond their national territory and in the direction of the other party. Furthermore, Article IX of the Outer Space Treaty calls

---

[14]See F. Kenneth Schwetje, "Considerations for Space Planners" in *Proceedings of the Tenth Aerospace Power Symposium: The Impact of Space on Aerospace Doctrine* (Maxwell Air Force Base, Alabama: Air War College, 1986), 20–6.

[15]For a discussion of Soviet attitudes toward space inspection see Malcolm Russell, "Soviet Legal Views on Military Space Activities" in Durch, *National Interests and the Military Use of Space*, 215–6.

on states party to the agreement to undertake consultations prior to any planned space activity or experiment that they believe may cause harmful interference with the activities of other states. On the national level, the North American Aerospace Defense Command (NORAD) also acts as a clearing house for all U.S. laser experiments that involve the projection of beams into outer space. It provides both warning to space system controllers and guidance to the laser operators. The acceptance of an agreement that stipulated advanced notification of all space weapons tests would again depend on whether either party felt that it would facilitate foreign intelligence gathering on the capabilities of the weapon systems being tested. Both superpowers have accepted the requirement of advance notification before testing some types of ballistic missiles, but this arrangement has nevertheless led to some friction over the monitoring and degree of access to test data.

Technical Requirements

Each of the potential rules of the road agreements discussed above would require monitoring equipment and other technical devices either to ensure compliance or generally facilitate their operation. These technical requirements are discussed as they apply to the various options.

A Prohibition on the Use of Force in Space. Parties to any agreement prohibiting the use of force in outer space would need to be able to distinguish between deliberate interference with their spacecraft and accidents or malfunctions that have the same effect. For example, a satellite collision with a piece of space junk is virtually indistinguishable from deliberate destruction by a kinetic energy weapon, but the two events clearly have very different implications. Moreover, parties to an agreement would need to be able to determine the source of the hostile act so as to place appropriate blame. Although the likelihood of warfare in space is almost always couched in terms of U.S.-Soviet interactions, the growing capabilities of other states to engage in hostile space activities should not be overlooked.

Because virtually all satellites possess some self-monitoring and diagnostic equipment that periodically reports information on their "state of health" to ground controllers, the United States and the Soviet Union already possess some means for determining the likely cause of spacecraft malfunctions. Other sensors could be added to provide further information, particularly on whether the malfunction was internally or externally induced. These include sensors to detect sudden accelerations following an impact with a solid object (accelerometers), laser irradiation (thermistors and light indicators), and damage from a nuclear explosion (radiation detectors).[16] Some of these devices may already have been deployed on certain U.S. spacecraft to detect hostile interference.

---

[16]OTA, *Anti-Satellite Weapons*, 110, 115.

These sensors would provide information on the nature of the interference with a satellite, but they might not always be able to determine whether this was the result of a deliberate act. The most pertinent question to ask following an unexplained incident in space would be whether it was an isolated event or part of a series. How important were the mission(s) performed by the affected satellite(s)? When and where did the incident(s) take place? Were other spacecraft or objects detected in the vicinity of the affected satellite(s) at the time of the incident? Was the satellite "illuminated" by radar or lasers prior to the incident occurring—suggesting that an attack was being planned or in progress? Were any objects launched, or lasers directed, into space at the time of the incident? Finally, did the incident occur at a time when hostile acts in space were plausible?

The answers to most of these questions would be provided by national technical means already deployed by the superpowers. These include space surveillance radars and optical devices, launch detection sensors, signals intelligence gathering equipment, and attack warning sensors that are being fitted to certain U.S. satellites. With expected improvements to U.S. and Soviet capabilities in the above areas as well as the deployment of sensors on satellites that can detect interference, both sides could verify an agreement prohibiting the use of force in space with high confidence.

An Incidents in Space Agreement. As outlined above, an incidents in space agreement would pose similar monitoring challenges to an accord that solely prohibited the use of force in space. The principal difference would be the greater emphasis placed on the tracking and identification of objects in space, the detection of space launches, and the propagation of directed energy beams into space. This new emphasis is necessary not only to reduce the likelihood of collisions in space but also to ensure compliance with the provisions banning dangerous maneuvers, simulated attacks, and the launching of objects in the direction of passing satellites.

Currently, the United States and the Soviet Union each operate an extensive network of ground-based radars, electro-optical telescopes, and other miscellaneous imaging systems to detect, track, and classify objects in space.[17] The collection of telemetry and command and control communications to and from active satellites by signals intelligence collectors provides additional information on the location and function of active space systems. Although the various ground-based sensors and other intelligence gathering equipment collectively provide impressive space

---

[17]See Appendix B of Stares, _Space and National Security,_ for more details on U.S. and Soviet space surveillance capabilities.

surveillance capabilities, they do have significant limitations.[18] Radars, for instance, are limited to line-of-sight observations leading to periods when space objects cannot be monitored. The reported instances when space surveillance radars have experienced difficulty in finding satellites that have maneuvered between observations illustrates the problem of keeping track of objects that can change their orbit. Objects below a certain size are also difficult to track. Moreover, the detection and tracking of objects at great distances from earth, particularly in the geosynchronous orbit, is not easy. The necessarily large radars are generally unwieldy for some surveillance tasks.

Even the latest generation of U.S. electro-optical sensors, the Ground-based Electro-Optical Deep Space Surveillance System (GEODSS), cannot provide highly detailed images of objects in the important geosynchronous orbit. They appear little more than small dots of light on the screens of the operators.[19] Nevertheless, this system can determine the basic characteristics of the object's orbit, such as whether it is stable, turning, or tumbling and its rate of movement. Moreover, it can do this rapidly. Each GEODSS sensor can differentiate satellites from stars and space junk in a single field of view within five to twenty seconds.[20] This information can in turn be relayed immediately to the North American Aerospace Defense Command for comparison with their log of space objects. The principal limitation of the system is that its observations can only take place at night and in clear weather.

The current level of space tracking has some shortcomings, but improvements can be expected in the future. New techniques are being actively explored in such areas as adaptive optics, inverse synthetic aperture radars, laser imaging radars, and interferometry.[21] The largest contribution to space surveillance, however, is likely to come from a constellation of space-based long-wave infrared (LWIR) sensors. With a moderate-sized constellation of four to six satellites, the level of space surveillance would

---

[18]Dennis K. Harden, "Current Capabilities and Future Requirements of the Air Force Space Surveillance Network," in U.S. Department of the Air Force, *Proceedings of the Tenth Aerospace Power Symposium: The Impact of Space on Aerospace Doctrine* (Maxwell Air Force Base, Alabama: Air War College, 1986), 54–64.

[19]Dale Foust, "GEODSS Update," *Quest*, vol. 6, no. 2 (Summer 1983); Anne Randolph, "USAF Upgrades Deep Space Coverage," *Aviation Week and Space Technology*, vol. 118, no. 9 (February 28, 1983): 57–8. "GEODSS Photographs Orbiting Satellite," *Aviation Week and Space Technology*, vol. 119, no. 22 (November 28, 1983): 146–7.

[20]Randolph, "USAF Upgrades Deep Space Coverage," 58. Each sensor also has an auxiliary telescope for tracking objects in low earth orbit.

[21]For a discussion of evolving monitoring technologies see Kosta Tsipis, David W. Hafemeister, and Penny Janeway, eds., *Arms Control Verification: The Technologies that Make It Possible* (Washington, D.C.: Brasseys-Pergamon, 1986).

be extended considerably.[22] Under the Space-Based Space Surveillance program the U.S. Air Force was pursuing the development and deployment of such a system. This program, however, has now become part of the SDI's Space Surveillance and Tracking System program. Although the envisaged system, if deployed, would improve U.S. space-tracking capabilities, it will not be optimized for this role. Rather, the primary mission will be to detect and track ballistic missile warheads during the mid-course phase of their flight. In addition to space-based infrared sensors, placing radars in space to cover such specific areas as the geosynchronous orbit is another possible option.

Many of the systems described above are also used to determine the basic functions of satellites by their externally observable characteristics and the signals they receive and transmit. Reconnaissance satellites can also be used to inspect objects in low earth orbit providing they can be maneuvered into the general vicinity and their sensors can be redirected accordingly.[23]

An agreement that prohibits the placement of weapons in space or in specific orbits would require that the function of every spacecraft be positively identified. Positive identification is not essential for monitoring compliance with an incidents in space agreement. Instead, the primary task would be to determine whether space objects were being maneuvered or deliberately used in a way that was hazardous to other spacecraft. Although some simulated attacks on spacecraft, such as close proximity fly-bys, might go undetected—certainly in areas where both the United States and the Soviet Union have limited space tracking coverage—placing attack warning sensors on spacecraft would reduce the probability of this occurring. The deployment of space-based infrared sensors discussed above also would reduce the scope for such covert activities and, furthermore, help detect the propagation of laser beams from ground sites and space systems that are being directed in an intimidating manner.[24] Finally, it is important to recognize that one of the principal functions of an incidents in space agreement is to prohibit brinkmanship and acts of harassment in space; such activities are pointless if they are carried out covertly or in a way that risks being undetected. Overall then, an incidents in space agreement can be adequately verified.

---

[22]*Department of Defense Authorization for Appropriations for Fiscal Year 1986*, Senate Committee on Armed Services, 99th Congress, 1st session. (Washington, D.C.: GPO, 1985), pt. 7, 4288. See also OTA, *Anti-Satellite Weapons*, 78–9, for a discussion on LWIR sensor technology.

[23]When the space shuttle *Columbia* shed some of its protective thermal tiles during its maiden flight, there was some speculation that a U.S. reconnaissance satellite was maneuvered specially to inspect the damage. See Anthony Kenden, "Was 'Columbia' Photographed by a KH-117?" *Journal of the British Interplanetary Society*, vol. 36 (1983): 73–7.

[24]OTA, *Anti-Satellite Weapons*, 110.

"Keep-Out" Self-Defense Zones. An agreement that delineated portions of space as self-defense zones also would require space surveillance systems to monitor potential encroachments and covert deployments. Although the identification of objects in the geosynchronous orbit is currently difficult, this procedure is not strictly necessary to ensure compliance. Any unauthorized entry by a space object into a prohibited zone would be suspect. It is difficult to believe that repeated encroachments and prolonged covert deployment would go undetected even with current monitoring capabilities.

Crisis Management Procedures. Establishing communication channels to exchange information in a timely fashion presents no insurmountable technical problems. The equipment used to establish the hot line and, more recently, nuclear risk reduction centers would be adequate for this task.

Cooperative Monitoring Procedures and Confidence-building Measures. An agreement stipulating prior notification of space launches would not require special or additional monitoring facilities. The United States and the Soviet Union currently use space-based infrared sensors to detect the launch of ballistic missiles and space boosters. The three U.S. launch detection satellites are deployed in the geosynchronous orbit, while the Soviets maintain—when fully operational—a constellation of nine satellites in a highly elliptical polar orbit commonly called a *Molniya* orbit after a class of communications satellites that are deployed in a similar way. In addition to these sensors, both superpowers employ ground-based radars to detect space launches once the booster rises above the horizon. The Soviets also have deployed radars with over-the-horizon detection capabilities.

Although no instances of the United States discovering a space object that went undetected at launch have been reported, U.S. monitoring capabilities can be expected to improve with the planned upgrades to the early warning satellites of the Defense Support Program. The Boost Surveillance and Tracking System being developed as part of the SDI would likewise improve U.S. capabilities, if it were ever deployed. These improvements, however, are intended principally to help characterize, rather than detect, space launches.

In addition to the use of reconnaissance satellites to inspect objects in space, the United States could also employ the space shuttle for this purpose at altitudes no higher than 600 kms. If developed, orbital maneuvering vehicles deployed from the shuttle payload bay could also be used for inspection purposes. In addition to optical surveillance, other inspection techniques (for example, infrared sensors and gamma ray spectrometers) would be feasible.

In summary, there are no insurmountable obstacles to verifying the rules of the road options outlined. Many of the functions necessary to observe compliance are already carried out by the United States and most probably by the Soviet Union. The main verification tasks in effect would represent "business as usual" for the superpowers. Nevertheless, monitoring capabilities in certain key areas, principally space surveillance, could be improved with space-based detecting and tracking devices.

Potential Spin-offs

The successful negotiation of a rules of the road agreement for space operations could have a number of useful spin-offs. In particular, it could provide impetus to the regulation of other areas of space activity. One example is an agreement to remove inactive satellites from key positions around the geosynchronous orbit. Similarly, nations could agree to reduce the proliferation of space junk by banning, for instance, the deliberate destruction of spacecraft in certain orbits. The safe disposal or removal of nuclear power sources in orbit is another potential area of agreement.

Rules of the road may also create a better climate for more cooperative ventures between space nations. These include scientific cooperation, information and personnel exchanges, and even joint missions for planetary exploration.

Evaluation

Rules of the road would not constrain the development of weapon systems in space, but they could help avoid or mitigate some of the undesirable consequences of their deployment. Of the potential agreements discussed above, several are worth pursuing. An incidents in space package of measures that included a prohibition on the use of force in space, as well as dangerous maneuvers and simulated attacks, would be particularly desirable if dedicated communication channels for crisis management purposes were also established. Prior notification of space launches and space weapon tests as well as the joint maintenance of a common catalog of objects in space could also be incorporated into such a package of measures. An agreement of this kind would not add to the space surveillance tasks of the superpowers; they already have an interest in monitoring activities in space and potential interference with their spacecraft. Nevertheless, both the United States and the Soviet Union would benefit from improvements to their current space surveillance capabilities.

An incidents in space package initially could be a subject for bilateral discussions between the superpowers. The U.S.-USSR Agreement on Cooperation in the Exploration and Use of Outer Space for Peaceful Purposes, signed on April 15, 1987, provides the basis for such discussions.[25] Once concluded, an agreement could then be opened up for multilateral endorsement in much the same way that the Outer Space Treaty was

---

[25]See text of the agreement reproduced in *Space Policy*, vol. 3, no. 4 (November 1987): 353–4.

negotiated and implemented under UN auspices. Alternatively, multilateral discussions could be held from the outset among the existing space powers. These would include representatives from the European Space Agency as well as Japan, China, and India—all of which have active space programs. A multilateral initiative, moreover, might be a more fruitful approach for achieving an agreement between the superpowers.

The other potential rules of the road discussed in this paper, however, do not appear to be attractive propositions at this stage. A self-defense zone agreement, for example, would probably be viewed by the Soviet Union as condoning space weapon deployments. If such deployments do become a common feature of space activities in the future, then self-defense zones may become more attractive. The same also applies to cooperative monitoring regimes in space.

In lieu of actual limitations on space weapons research and development, a rules of the road agreement represents the next best option. Although the benefits are relatively modest, an incidents in space agreement is a worthwhile endeavor with little cost or risk to the parties involved.

7

TECHNOLOGY AND INTERNATIONAL PEACEKEEPING FORCES

BY WILLIAM H. LEWIS

Technology has become the holy grail of late twentieth-century global politics. It holds out the hope of solutions to many otherwise intractable problems ranging from declining agricultural growth rates to mounting birth rates, from the trials of nation building to the tribulations of order building, and from the conquest of space to the violent resolution of disputes among nations. Paradoxically, policymakers and research specialists have devoted little attention to the potential of technology for enhancing international peacekeeping.

Part of the reason for this omission can be attributed to the fact that, unlike domestic politics, relations among nations take place in an arena that has no central governing body. No agency exists at the supranational level with the authority and resources to impose settlements. Although nations can enter into treaties and contract obligations voluntarily, no effective body exists to insure compliance or to impose effective sanctions against violators of international norms and conventions. Nevertheless, there is growing recognition of the need to bolster the capacity of the world community to resolve disputes peacefully. These questions must then be asked: What institutions are available for peacekeeping duty and what role can the United States assume—directly or indirectly—in supporting international peacekeeping initiatives?

Peacekeeping obviously is no panacea for interstate disputes. It cannot transform basic relationships between adversaries, particularly when they reflect cultural and ideological differences, or when national leaders view issues in zero-sum terms. Nor is peacekeeping a viable instrument for arms control. When rivalries are deepseated, they often are expressed in terms of a search for comparative advantage through arms acquisition. On the other hand, traditional peacekeeping operations, in small but important ways, can serve to limit the scope and the destruction produced by armed conflict as well as to provide broad foundations for confidence building in the cause of peace.

Despite arguable results, the United States has been committed to peacekeeping under international auspices at least since the formation of the United Nations (UN). Successive administrations have extended financial support to multinational efforts in the Middle East, Africa, and Asia. Rescue and relief operations also have received material U.S. backing, such as that provided during the Shaba crises of 1977 and 1978 and the Organization of African Unity (OAU) intervention in Chad at the beginning of this decade. Current U.S. policy calls for the buttressing of UN peacekeeping capabilities through the earmarking by member states of national contingents for use by the secretary general. The Reagan administration also has indicated that it intends to make U.S. airlift capabilities, technical expertise, and military equipment available for such purposes. Furthermore, it has proposed to the Congress that a Special Peacekeeping Fund be established within the framework of the security assistance program. Congress has approved the fund, but within the parochial framework established by the administration—that is, to fulfill U.S. obligations to help fund UN "forces" assigned to southern Lebanon and Cyprus.

The United States has been laggard in making its advanced technology available for multilateral peacekeeping operations. Agencies within the executive branch of the U.S. government assigned primary responsibility for planning logistical support for international peacekeeping have failed to agree upon an action program for stockpiling and using advanced technologies in the event of need by the UN. What contact has taken place with the UN headquarters staff has been intermittent and unproductive. Reflective of this deficiency was the inability of the United States to respond to the Norwegian request for assistance for their United Nations Interim Forces in Lebanon (UNIFIL) contingent in 1978, a failure that proved embarrassing to the Carter administration.

On the bilateral level, however, technology played an important role in U.S. efforts to separate Israeli and Egyptian forces in the Sinai Peninsula during the 1970s. It helped thereby to lay the foundations for the Camp David peace negotiation that finally terminated the more than thirty-year state of war between the two countries. Placed at the disposal of peacekeepers in the Sinai was a sophisticated complex of acoustic surveillance devices and communications equipment that permitted American civilian personnel to detect movements by Egyptian or Israeli military units toward or through the demilitarized zone separating their two armies. The U.S. role, as Vice President Walter F. Mondale declaimed, was to act as the "eyes and ears of peace."

This chapter identifies the ways in which technology could be used to help preserve the peace, despite the new and yet more complex threats to peace likely to emerge over the coming decade. It outlines several types of crises that could evolve in Third World regions, provides general parameters for the type of peacekeeping requirements that would arise in each, and then relates these requirements to existing or emergent advanced technologies, military and civilian, that can enhance future peacekeeping operations. Costs are briefly evaluated, as well as technical requirements for training, operations, and maintenance. In addition, local sensitivities in deploying cutting-edge technologies and questions relating to protecting classified equipment from third parties are taken into account in this discussion.

Although this chapter does not provide an exhaustive analysis of the relationship of technology to peacekeeping, it does conclude that technology can contribute to more effective peacekeeping operations. Given the likely nature of future conflict situations in the developing world, a clear imperative to upgrade the technological capabilities of future peacekeeping operations is incontestable. On the other hand, it is clear that enhanced capabilities present attendant difficulties. For example, the deployment of surveillance equipment with great range and intrusive capabilities might prove unacceptable to local disputants under certain circumstances. At the same time, the denial of such equipment to peacekeeping forces might well endanger their security and effectiveness. In short, advanced technology can be an asset or a liability depending on local conditions and the goals of contending parties.

The Need for a Policy Change

Technology offers a panoply of instruments to serve U.S. foreign policy goals and objectives. Successive administrations have pledged to make these instruments available in the cause of international peace and order. Rhetoric aside, however, Washington has developed no clear guidelines as to when and where U.S. technical capabilities will be made available to support international efforts at conflict resolution, the degree to which the United States will support international organizations in their efforts to plan for peacekeeping, or what type of collaborative effort the United States is prepared to organize, or to join, in stockpiling equipment and other resources that might be required to meet a wide range of peacekeeping operations.

Successive U.S. administrations have preferred to "keep options open" when addressing Third World conflict situations. The traditional matrix of superpower rivalries—encompassing balance of power, spheres of influence, and zero-sum considerations—all too frequently have frustrated efforts at a more balanced evaluation of policy alternatives and choices. Within this context, perceptions of the power, prestige, and influence expected to be derived if the conflict situation is resolved on terms favorable to friends and allies of the United States frequently proves overriding. Within the vortex of policy, the capacity of the UN and regional organizations to stabilize local conflicts, and to terminate them on terms noninjurious to the contending parties, is accorded only marginal consideration by most policymakers. Only in extremis, and after extensive domestic debate within congressional precincts, do senior echelons within the U.S. executive branch look to external sources for peacekeeping purposes.

In short, public declarations to the contrary, there is only a feeble impulse within the U.S. government to adopt policies, plans, and programs to enhance the peacekeeping capabilities of international organizations. What little attention is devoted to the subject is episodic, lacking in conceptual clarity, and deficient in terms of meaningful application.

Perhaps nothing so clearly illustrates the point as the difficulties experienced by the Reagan administration in dealing with the Persian Gulf conflict in the wake of the tragic circumstances surrounding the attack on the USS *Stark* in mid-May 1987. The attack on the *Stark* transformed the terms of debate in the United States relating to basic U.S. interests in the Persian Gulf. According to statements by various executive branch officials, in providing armed escort to "Kuwaiti-U.S." flag vessels the United States has taken as its basic goal one or more of the following aims: (a) asserting the right of free and unfettered navigation (in an admitted war zone); (b) insuring access to the region's oil supplies; (c) frustrating Soviet and Iranian ambitions in the region; (d) reversing the downward spiral in America's reputation among "moderate" Arab states in the wake of the Irangate affair; or (e) aligning the United States with Iraq in its hostilities with Iran in an effort to insure a military stalemate, and thereby foster a cease-fire and negotiated settlement of the seven-year-old conflict.

Although one or several of these goals may have been unexceptionable from the perspective of U.S. policymakers and practitioners, they placed Washington at the center of the conflict between Iran and Iraq. More critically—the risks of costly collisions with Iran aside—the Reagan administration's approach raised serious questions concerning the professed position of U.S. neutrality in the war.

The intervention by the United States in the Gulf conflict has provoked renewed public debate over the circumstances in which the United States should commit its military forces in distant hot spots. As one commentator has observed:

> An important measure of the readiness of any 16 or 18 people seeking the presidency is his capacity to define the standards he would use in deciding whether to deploy military force or intervene militarily in a world hot spot. Equally important, given the continuing domestic conflict over this category of issues, is a clear statement of the steps he is prepared to take to involve Congress in such a decision and the limits he would place on the congressional role.
>
> The casualties we have suffered—and those inevitably to come—compel a serious national debate on these issues during this election cycle.[1]

---

[1]David S. Broder, "How to Decide When to Use Force," *Washington Post,* May 27, 1987, A–21.

Particularly troubling has been the inclination of several administrations to intervene unilaterally in distant crisis situations. Almost invariably, the majority of this nation's traditional friends and allies tend to reject requests for military collaboration of the type experienced during the Korean War, opting instead for traditional diplomacy or even outright passivism. As a result, the American public, not without reason, concludes that the United States has few takers in sharing risks associated with the injection of military forces into crisis situations. A sense of estrangement has begun to crystallize in the public mind, and with it a perception that the United States has few reliable partners in our endeavors to protect mutual interests and to restore some semblance of order in a disorderly world. The French refusal to support the U.S. retaliatory attacks against Libya in 1986 in the wake of Libyan-inspired terrorist attacks against American military personnel in Europe is but the most memorable of many examples of our allies' inconstancy.

It is not surprising that burdensharing has become the leitmotif of those who wish to lighten the self-imposed onus that the United States has carried for four decades. Yet we too must bear part of the blame for our allies' intransigence. For while Washington has sought support from Japan and the NATO allies in the Persian Gulf, in southern Africa the United States has conducted its own diplomatic initiatives without meaningful prior consultations with these same allies. And in Angola the United States has been joined only by South Africa in providing "covert" aid to the UNITA movement and sustaining a guerrilla war that has been under way for more than a decade.

Technology and Conflict Situations

The Third World has begun to approach systemic overload as numerous conflict situations have emerged over the past decade, and most of these have not been resolved. Some conflicts simply do not lend themselves to early resolution. In Angola, Ethiopia, and Sudan, for example, great power interventions intensify and prolong tribal, ethnic, and cultural conflicts by stifling discussion and compromise among the contending indigenous parties. And yet some conflicts may be amenable to resolution through intercession by international organizations prepared to serve as honest brokers in arranging cease-fires, monitoring these arrangements, or making peacekeeping contingents available to ensure compliance with the terms of agreement. Advanced technology in the hands of the peacekeepers could help to improve the effectiveness of their performance. Only a few cases are needed to illustrate the potentialities of existing technologies.

The United Nations Peace-Keeping Forces in Cyprus (UNFICYP), initially deployed in 1964 to preserve law and order, continue to play an important, although somewhat truncated, intermediary role. The national

units assigned to UNFICYP would find their burdens eased if more advanced command-and-control and communications equipment were
placed at their disposal. Such equipment is readily available in U.S. Department of Defense stocks, but would require release through excess need
declarations by the Pentagon—a procedure that the present administration is loathe to pursue.

In Lebanon, similarly, UNIFIL confronts daunting problems in the face
of a deteriorating security situation in southern Lebanon. In order to provide for its own self-defense, as well as protect the communities in its patrol zone that are subject to harassment and occasional armed attack by
contending local forces, UNIFIL will require significant improvement in
the monitoring, observation, and communications capabilities of its various national units. Once again, the United States and other West European nations can readily provide equipment upgrades if their governments
approve such a course of action.

Outside the framework of existing UN operations, a wide range of
potential peacekeeping requirements can be discerned. These need not
all be met under the auspices of the United Nations. Regional organizations, multilateral groupings, or even private entities may serve as useful
sources of assistance.

Over the next several years, peacekeeping opportunities are certain
to arise in regions where illegal border crossings serve as a catalyst for
instability and local conflict. Among the prime candidates for international interdiction or control are: the area adjacent to the Jordan River,
the natural boundary between the West Bank and east Jordan, where an
integrated warning system parallel with the river could provide Israel with
enhanced tactical alert capabilities and thereby, perhaps, facilitate the creation of a Palestinian entity; the poorly demarcated Ethiopia-Somalia border region, where the Somalis, already the victim of invasion by Ethiopian
forces in 1982–83, clearly require early warning capabilities to turn aside
future "invasions"; the Thai-Kampuchean frontier, where the pursuit of
Khmer guerrillas by Vietnamese forces has produced sporadic collisions
between the armed forces of both countries; the Angola-Namibia border,
where various forces continually cross to raid adversary military formations and where any settlement of the Namibia problem under UN
auspices will require border surveillance and control sponsored by the
UN or another third party; Central America, where countries that share
borders with Nicaragua would undoubtedly have greater confidence in
a peace settlement sponsored by the Contadora nations if effective border
controls and monitoring capabilities were available.

The Third World is an area where a wide range of conflict situations are
certain to arise and where efforts to mute disputes, to stabilize local border situations, and to monitor compliance with the terms of internationally
sanctioned agreements will place increased burdens on organizations with

peacekeeping responsibilities. The primary requirement in the years immediately ahead will be to match responsibilities with capabilities.

Surveillance Missions in Peacekeeping Operations

The idea of using technical means of surveillance to enhance peacekeeping capabilities has been under discussion since the mid-1960s. Recent developments in electronics and continued demand for greater security, both private and national, have increased both the supply and demand of new technologies for area surveillance. Television systems, remotely piloted aircraft, and intruder detection systems have the greatest potential utility for peacekeeping operations.

Television systems are, of course, familiar in the United States as means of remotely monitoring apartment buildings, offices, hospitals, banks, and so on. They can be emplaced in nearly any environment with sufficient field of view (they would be worthless in a jungle, for example, but excellent in a desert). They can operate at night and in bad weather with infrared sensitive tubes. They can be fitted with telescopic lenses, remotely aimed, and camouflaged or armored against stray bullets. They need not be watched continuously by on-duty personnel; they can be "watched" by computers designed to summon monitoring personnel in response to certain external stimuli. And remote television sensors and scanning systems are well developed. The "Total Scan Surveillance System" produced by the Western Bureau of Investigation, for example, links up to fifty television cameras into a centralized computer console. A memory disc handles routine surveillance by telling the rest of the network what ought to appear in a particular scan. If no discrepancies are observed between the actual scan and the record on the memory disc, then the system continues its operation. If some discrepancy occurs, however, an alert sequence is triggered; an alarm is raised and visual display units give enforcement personnel an instant replay to pinpoint events before, during, and after the infringement took place.

Other, more sophisticated, devices are available to detect intruders into a controlled area. Complex integrated remote systems were employed in the Vietnam-era McNamara Line and were even used experimentally by the French in Algeria. During the Vietnam conflict the U.S. Army developed a full family of ground-based radars and other sensors to detect surface movements at night or in bad weather. A typical system (still in inventory) is the AN/PPS-5, a man-packed, air-droppable or vehicle-transported radar system designed to detect, locate, and identify enemy personnel and vehicular targets out to line-of-sight ranges of 6,000 to 10,000 meters. The system can be assembled in less than ten minutes. It is powered by a fifty-watt rechargeable battery or external source and can be remotely operated up to fifty feet from the basic radar set. It provides an automatic sector azimuth-scanning and range-covering capability with both a visual and audio display of moving targets.

Such radars constituted only one element of the multisensor systems developed in Vietnam. Sensor devices positioned either through the air or by hand were used to detect the movement of humans within a range of about 40 meters and of vehicles within 300 meters. Ground surveillance radars and night observation devices were also employed successfully to identify potential targets. The radar sets organic to division maneuver battalions were used primarily to provide short- and medium-range limited visibility. Along with the radars there were night observation devices, either the older infrared lights or the newer starlight scopes. These sensors were of limited value in themselves, but when properly integrated into an overall plan proved to be most effective.

Development of these integrated systems has continued. Currently available types can utilize vibration, ultrasonic sound, infrared, or microwave radiation as the medium of detection. Whatever the medium employed, the system remains passive as long as conditions in the protected zone are constant. If an intrusion occurs, however, the protective web is shaken and alarms are engaged that may then trigger other systems into action. The current trend is to aim for fail-safe mechanisms that utilize more than one mode of detection to make them less prone to false alarms.

In terms of mixed, remotely monitored surveillance systems, the state of the art is probably the U.S. Army's Remotely Monitored Battlefield Sensor System (REMBASS). REMBASS is an unattended, ground sensor system that will detect and classify intruding personnel and vehicles. REMBASS utilizes remotely monitored sensors emplaced along likely routes of enemy approach. These sensors respond to seismic, acoustic, infrared, or other mechanical energy and magnetic field changes to detect enemy activities. The seismic sensor also identifies an intruder as a wheeled or a tracked vehicle or as a person. This sensor information is incorporated in short digital messages and transmitted by a sensor-contained FM radio transmitter. The sensor communicates with the REMBASS monitoring set, either directly or through repeaters. Messages at the receiver are demodulated, decoded, temporarily displayed, and recorded to provide a time-phased record of enemy activity. REMBASS utilizes seven types of sensors that can be emplaced by hand, aircraft, or artillery.

Remotely piloted vehicles (RPVs) can also aid area surveillance operations and have already been used in Israel. Remotely piloted or remotely directed aircraft are typically small and made of non-radar reflective materials; they can neither be seen visually nor be detected on small radars. Loitering above areas of interest, they can use cameras, television, infrared, and other sensors for surveillance. Their advantages over conventional aircraft are much lower cost and reduced pilot risk.

The most ambitious RPV program is the U.S. Army's Aguila system. It is, however, not yet fully field tested and approved for production. Moreover, it will probably prove to be too advanced and expensive for use by peacekeeping forces, due to its incorporation of jam-resistant controls and high-technology target designation equipment to guide weapons. Less complex systems, such as Pave Tiger or Skyeye, or existing Israeli systems, may be better suited to the pure surveillance role.

Other Peacekeeping Roles

In addition to surveillance, peacekeeping missions have been assigned two other functions. These require interposition of truce supervisory forces between warring factions and, when circumstances dictate, separation of the factions. A variety of local and other factors have to be taken into account before possible applications of technologies in these roles can be considered. Among the more salient are:

- the financial resources and technical capabilities of nations willing to support peacekeeping operations with national contingents;
- the restraints and restrictions placed on peacekeeping forces by their sponsors, local adversaries, or third parties;
- the missions, roles, and other responsibilities levied on these forces; and
- other constraints such as deadlines for mission completion, environmental factors, traditional support requirements (engineering, communications, medical), and related considerations.

Consideration of these factors leads, inescapably, to the conclusion that applications of technology must be weighed on a case-by-case basis. This applies with particular force to the introduction of advanced systems into peacekeeping operations where the absorptive capacities of the designated forces are attenuated or where the viability of the assigned operation is doubtful.

The increasing swiftness of modern conflict, moreover, especially between technologically advanced states, suggests that peacekeeping forces may have to be emplaced more rapidly than in times past if they are to be effective. Modern logistics and transport technologies can facilitate this. They also offer the possibility of lower delivery costs, particularly in roadless or rugged terrain and other harsh environments.

The advantages of modern logistics and transportation can most effectively be realized if governments and multilateral organizations are prepared to employ them swiftly. But effective exploitation of innovative techniques and technologies places a premium on prior planning and efficient logistical support—requirements that are infrequently met in most multilateral peacekeeping efforts. The United States traditionally has been at the forefront of nations supplying airlift and other forms of logistical support for UN and other multinational peacekeeping and relief operations, including those in the Congo (1960), Yemen (1973), Shaba (1978), and Chad (1979 and 1986), and the American government could take the lead in organizing and underwriting the planning by international organizations necessary to facilitate the adoption of modern logistics and transportation.

The separation function is accomplished primarily by spatial "distancing" of hostile forces along a demilitarized zone or line that is patrolled or otherwise monitored by peacekeeping forces. Traditionally, this function has been carried out by foot patrols; but it could be carried out by remote sensors, monitored at central locations. Remote sensors make possible the use of smaller and less costly forces, and can minimize friction between peacekeepers and the forces they seek to separate. Systems do not tire or become less alert because of previous false alarms or inaction. And there may be political advantages associated with remote means of surveillance resulting from the reduced visibility of peacekeepers.

The experience of the U.S. Sinai Field Mission in the 1970s is particularly instructive. The mission acted as an integral part of the overall early-warning arrangements agreed to in September 1975 by the governments of Egypt and Israel. That agreement—the Sinai II Agreement—capped almost two years of intensive negotiations following the October 1973 war. An especially difficult problem had been the reconciliation of Egypt's insistence on full Israeli withdrawal from the strategic Mitla and Giddi Pass areas in western Sinai with Israel's conviction that to retain its extensive electronic surveillance facility atop the escarpment near the western entrance to the Giddi Pass was essential to its security. The impasse was resolved when Secretary of State Henry A. Kissinger secured agreement in Cairo and Jerusalem to the establishment of a buffer zone in which the United States would mount and operate tactical surveillance of the two passes as part of overall early-warning arrangements. The tactical warning system, according to the terms of the agreement, would be staffed by civilian volunteers, would be located in a demilitarized buffer zone, would be controlled by the United Nations, and would separate two belligerent armies.

The agreement rested on five basic elements: the political decisions of Egypt and Israel to avoid further hostilities and to reduce the chances of inadvertent conflict; the physical separation of the two armed forces and their willingness to limit the numerical size and capabilities of their forces in areas contiguous to the buffer zone; the presence of neutral—UN and U.S.—personnel to monitor compliance with the terms of the agreement; aerial reconnaissance flights along the median line of the buffer zone by the United States, Egypt, and Israel; and strategic surveillance by visual and electronic means, conducted by Egyptian and Israeli personnel, together with early tactical warning by the United States to avoid surprise attack and minimize unwarranted suspicion.

The tactical early-warning system deployed by the United States performed essentially a route surveillance function keeping watch over the

Giddi and Mitla Passes. The surveillance system established by the Sinai Field Mission included unattended ground sensors—four sensor fields in all—and visual coverage by civilian personnel from three watch stations.

All sensors performed the same basic function. They transmitted a radio signal to an attendant watch station whenever a moving object entered their respective sensing zones. A mixture of sensor types was used in the early-warning area because of variations in environment and soil conditions. For example, in some sectors seismic sensors were selected because the geology enabled seismic sensors to detect personnel and vehicles at long ranges, thus requiring a minimum of sensors to cover a particular mountain pass. In areas where the geology limited the detection range of seismic sensors, infrared or strain sensitive sensors were selected. And in a few instances, seismically activated microphones were utilized. When watch station operators were not able to identify movements in a sensor field, UN patrols were dispatched to investigate and take whatever steps were necessary.

In the years since the operations of the Sinai Field Mission were concluded, several new generations of ground monitors have emerged that enhance capabilities for route, border, and general area surveillance. Virtually all involve sensors based on the detection principles of seismic, acoustic, infrared, magnetic, electro-magnetic, pressure, electric, and earth-strain disturbances. Optical and electro-optical devices for improved day and night visibility also have been developed. All of these surveillance technologies are currently available in the United States and could be adapted to enhance multilateral peacekeeping efforts.

Maritime Surveillance

An equally pressing mission, of particular relevance today, is maritime peacekeeping. Armed conflict between nations, as in the Persian Gulf, can have adverse consequences not only for neighboring nations, but also for distant countries heavily dependent on the resources of the conflict-ridden region. To the extent that the warring nations threaten to disrupt the commerce and trade of neutral parties, or seek to close maritime chokepoints, the threat of third-party involvement is likely to grow. In such circumstances, multilateral peacekeeping forces could help to defuse crisis situations by maintaining the integrity of waterways.

The United States has a special interest, as a maritime nation, in preserving free, safe, and unrestricted transit of international waterways, and has entered into agreements and treaties intended to ensure safe passage and unfettered access to important trading partners. The U.S. government has also concluded numerous bilateral agreements on cooperative search and rescue operations, is party to the International Convention on Maritime Search and Rescue, and serves as a key member of the International Maritime Organization—a specialized agency of the United Nations formed in 1958, inter alia, to facilitate international trade.

As part of the U.S. effort to enhance international commerce, the government conducts extensive research and development activities to aid navigation in various waterways. For example, in 1985, the United States developed and began to install an "Automated Position System"—a microcomputer-based system installed aboard buoy tenders to aid the positioning, guidance, and reporting of ships. The system is an important tool for safe navigation in narrow waterways. Research is under way to improve capability for port clearance, tracking of ships in stormy ocean areas, and emergency communications for rescue operations—all of which should aid international commerce.

Threats to international commerce have taken various forms. As the number of incidents of covert minings of harbors and waterways, hijackings of vessels, piracy, and episodic acts of terrorism has increased in recent years, proposals for negotiation of new "laws" covering piracy and acts of terrorism have proliferated. Whatever the conclusions of the international community, however, enforcement measures will be required, both to deal with maritime "law and order" problems and to ensure free and unfettered passage of the vessels by neutral nations in "conflicted zones."

The *Stark* incident raises this question: Which of the following approaches is the most promising for safeguarding international commerce— unilateral action by one or two nations (as in the Gulf, with the United States and the Soviet Union operating separately and with competitive objectives in mind), a consortium consisting of the major maritime nations (including the United States and the Soviet Union), or a grouping of maritime nations (excluding the United States and the Soviet Union) operating under the auspices of the United Nations?

In many cases, conventional naval forces can be interposed for maritime peacekeeping. But, as witnessed in the Persian Gulf, two processes are gradually reducing the usefulness of traditional "gunboat" operations against nontraditional threats at sea. The first is the proliferation of cheap, effective antiship missiles in littoral states, especially those near such chokepoints as Hormuz, Malacca, and Suez. Traditional naval forces are steadily being barred from close, inshore operations unless accompanied by major air, antimissile, and electronic protection, preparations indistinguishable from those for wartime naval missions against major opponents. Under such circumstances, it is difficult to envision major naval powers such as France or the United States committing huge fleets to peacekeeping efforts; and volatile littoral governments cannot be expected to contemplate calmly the assembly of fleet units offshore on ostensibly peaceful pretexts.

A second factor is the increasing technological sophistication of the threat itself. It is true that Malacca-style pirates can be suppressed by conventional naval forces. But government-supported "pirates" equipped with intelligent mines, submersibles (such as Libya's six Yugoslav-built "Mala"– class midget submarines), and hand-held missiles may actually operate inside the threat detection envelopes of warships whose sensors and weapons have been optimized to deal with those of other major powers. The extreme difficulty that Sweden's naval and coast-defense forces encountered in locating small submarine intruders in the Hörsfjärden in the fall of 1982 is pertinent; six submersibles of unknown nationality were detected, but none were destroyed or even forced to the surface.

As experience in the Red Sea–Persian Gulf area attests, maritime surveillance and sea-lane maintenance can be a daunting requirement if conducted unilaterally. Although a number of assets are currently available to the United States, the range and sophistication of threats posed to American naval forces place heavy burdens on a single actor dedicated to defense of the principle of free navigation. The United States must bring to bear a full panoply of capabilities—ranging from satellite and aircraft surveillance, communications and early warning, mine demolition, surface and air defense, and retaliatory military power—which ineluctably adds to pressures to secure military access rights in contiguous nations. In addition, as the nature and scope of the threat widens, the United States is compelled, almost inevitably, to seek the support of other Western nations to share the attendant burdens and risks.

A more equitable and effective planning approach would be to establish institutional foundations for burdensharing within the framework of UN collective security responsibilities. Article 43 of the United Nations Charter envisages agreements between the UN and member governments on the provision of standby forces. A reconstituted Military Staff Committee should be encouraged to develop a peacekeeping maritime surveillance plan that would bring together standby forces with the capacity to conduct surveillance, protective free passage, mine clearing, and related operations. Countries such as Italy, the Netherlands, the United Kingdom, and West Germany all have substantial resources in this area and, with appropriate encouragement by the secretary general, could earmark these national resources as the nucleus for a standby maritime peacekeeping reserve.

Organizational Imperatives

New applications of technology that might enhance peacekeeping operations are under study or in the process of development. Factors of cost, complexity, and sustainability come into play when considering the feasible applications of this emerging technology. In addition, the composition of peacekeeping forces must be taken into account. A direct correlation between complex technologies and force capabilities is apparent— the more advanced the equipment being used in a peacekeeping operation, the higher the level of technical competence demanded of its operators.

The question thus arises whether the United States should continue to rely on the United Nations and regional organizations for peacekeeping operations in the trouble spots of the Third World or, instead, depend to a still greater extent on its own national resources and those of friends and allies.

From a cost-benefit perspective, unilateral intervention by the United States carries with it a number of potential disadvantages. At home, public and congressional opinion may prove unsupportive, and international opinion is likely to be volatile at best. U.S. military involvement in local conflict situations for peacekeeping is usually viewed with suspicion by Third World representatives, primarily on the grounds that big-power altruism is rare. Unilateral intervention undermines the potential of the United Nations and regional organizations to control conflict situations, fuels East-West competition to the disadvantage of local disputants, and threatens to widen and worsen conflict situations as other parties are drawn into the dispute.

The arguments in favor of continued recourse to the United Nations and regional organizations are many and compelling. First, despite its shortcomings, the UN has proved to be an important instrument for resolution of many disputes. The UN maintains an experienced peacekeeping organization available to intercede with little advance notification. And the prestige of the secretary general and the UN itself carries considerable weight in efforts to resolve many Third World conflicts. The secretary general can tap a large reservoir of nations able and prepared to contribute contingents to peacekeeping operations, thus ensuring a more equitable distribution of burdens. Finally, in theory at least, the UN role can serve as a buffer against big-power rivalries.

Offsetting these advantages, however, are several important liabilities. The veto power of the great powers in the Security Council and the dominance of the Non-Aligned Movement member states in the General Assembly often paralyze the UN. Funding for peacekeeping perennially runs short, thus undermining the credibility, authority, and effectiveness of the organization. The emphasis of the UN on decisionmaking through consultation and consensus building, moreover, consumes precious time and diminishes the organization's flexibility. And the diverse composition of most UN peacekeeping forces tends to engender monumental logistical headaches.

Strengthening UN Peacekeeping Forces

Despite the UN Charter's requirements that member states of the United Nations furnish standby forces for use by the Security Council, the members have preferred to make contingents available on a voluntary, ad hoc basis. Even when made available, they can be withdrawn without prior approval by the Security Council or the secretary general. This ad hoc approach to peacekeeping has produced a degree of "immobilism" within the UN Secretariat and the secretary general's planning staff, reflected today in the inability of UN members to agree even on the desirability of a permanent peacekeeping policy staff. UN personnel do little if any contingency planning and virtually no stockpiling of even the most basic equipment needed for peacekeeping.

In a report to the president on "Reform and Restructuring of the United Nations Systems," dated February 28, 1978, Secretary of State Cyrus Vance presented a number of constructive proposals for U.S. initiatives to buttress UN peacekeeping capabilities. Among the most notable were a suggestion that a body of highly trained military personnel ready to be called on short notice for peacekeeping and observation assignments be organized and funded by the United Nations and a recommendation that a UN "Peacekeeping Reserve" composed of national contingents trained in peacekeeping functions be created. The reserve contingents might be either combat or logistics units and should be available for service on short notice. The secretary also observed that member states not in a position to provide military units might earmark other facilities, presumably those which would enhance the surveillance and communications capabilities of the peacekeeping forces. And he urged that the United States declare its willingness to assist with "airlift of troops and equipment required for establishing a UN peacekeeping force authorized by the Security Council"—perhaps without reimbursement by the UN. Finally, the secretary of state recommended that the United States be prepared to examine with the UN "possible ways of upgrading the technical equipment available to observer missions and peacekeeping forces" and that ways of enhancing the observation and communications of UN forces through the use of modern technologies be explored.

An informal commission on disarmament and security issues organized by former Swedish Prime Minister Olof Palme in 1980 issued a report two years later containing several promising proposals for enhancing the peacekeeping capabilities of the UN and regional organizations.[2] The more promising proposals were: that the secretary general report to the Security Council on a regular basis throughout each year and present an annual "state of the international community" message to the Security Council at the foreign minister level; that a first step toward collective security in Third World conflicts arising out of border disputes be implemented by developing standard procedures for fact finding, military observation, and introduction of UN forces based on Third World support, a political "concordat" among the veto powers, and the availability of standby forces; that UN peacekeeping capabilities be improved by adopting standard training manuals, increasing assistance to Third World

---

[2]Olof Palme, et al., *Common Security* (New York: Random House, 1982).

countries in training and equipment, establishing regional arrangements for standby forces, stockpiling equipment, and earmarking special units for peacekeeping; that an agreement be reached on appropriate, regular funding mechanisms for peacekeeping; and that regional conferences on security and cooperation be convened.

Many of these proposals and recommendations warrant serious consideration and support. But realism suggests that, within the confines of the UN, modest building-block initiatives are prerequisites for an effective approach. A peacekeeping "system" already exists in theory, but it depends on a UN defense staff smaller than that of Gabon—four officers in New York, of whom the most senior is a brigadier—despite the existence of large UN operations in Cyprus, Lebanon, and on the Israel-Jordan border. A major push by Security Council veto members and the secretary general is needed to establish effective personnel planning and logistic support capabilities if the UN role is to be enhanced.

Possible New Initiatives

Since 1978 the United States has registered little significant progress in efforts to integrate emerging technologies with international peacekeeping requirements. Discussions with UN officials have been conducted intermittently and there are few signs that broad agreement has been achieved. Nevertheless, several fresh initiatives might be encouraged. These relate both to the UN and to the restructuring of U.S. resources in support of peacekeeping.

Within the UN, private resources and capabilities have not been fully exploited. A private organization, the International Peace Academy, formed in 1970, has conducted a wide array of seminars for policy planners and military, diplomatic, and other officials. Located in New York, the academy seeks to broaden the knowledge and understanding of participants from various nations of the skills and procedures that are required for peaceful settlement of international disputes (peacekeeping, mediation, negotiation, and so forth). Greater attention might be devoted to the role of technology in future seminars conducted by the academy. Because the academy is dependent wholly on private subventions, the U.S. government might seek to buttress its efforts by encouraging donations from the private sector, seeking the establishment of specialized training seminars, and working for the appointment of science and technology staffs to undertake specialized studies relating to advanced technologies and peacekeeping.

Given the many confrontational situations that exist in the world today, the need for surveillance systems employing advanced technologies is likely to increase. Various locations in the Middle East, Africa, and Latin America (particularly Central America) are good candidates for the emplacement of tactical surveillance systems in support of peacekeeping forces. Such systems, fixed or mobile, might incorporate photography, television, radar, thermal imaging, or light amplification technologies. Systems with magnetic, seismic, acoustic, or pressure-sensitive sensors that are either already in existence or under development must be adapted for peacekeeping purposes and integrated into supporting command and control, communications, and other informational arrangements. No less important would be the development of a range of surveillance technology systems that can be applied to a variety of different local circumstances.

The successful application of surveillance technology to peacekeeping operations, thus, will require rules, processes, and organizational structures for distributing surveillance capabilities and results from those capabilities. A regime for applying surveillance technology to peacekeeping might promote the joint use of facilities or capabilities by parties involved in the conflict and establish rules under which joint use would take place. The International Peace Academy, subsidized by private U.S. sources, should be encouraged to focus its future research efforts in the realm of surveillance technology applications.

A more ambitious undertaking, recommended by Robert C. Johansen, would be the formation of an international monitoring agency. In his words,

> Most governments already recognize that such an agency would aid multilateral peacekeeping and mutual security. More than 120 nations have already supported a modest French proposal to create a permanent international satellite monitoring agency. . . . Both the United States and the Soviet Union have opposed the French plan, although the Soviet Union's attitude toward monitoring. . . seems to be changing. A General Assembly study concluded that an international satellite monitoring agency was easily within technical reach. The start up and operating costs would be well under 1 percent of world military expenditures annually.[3]

If the UN is not in a position to create an international agency, other avenues might be explored. The secretary general might consider leasing communications satellite capabilities and could acquire remote-sensing data from the U.S. Landsat and French SPOT systems.[4] Reportedly, the Swedish Research Council is studying an alternative approach that would permit nonaligned countries to conduct monitoring with the assistance of European countries and Japan in launching monitoring satellites.

---

[3]Robert C. Johansen, "The Reagan Administration and the UN," *World Policy Journal,* vol. 3, no. 4 (Fall 1986): 279.
[4]Ibid.

Finally, within the UN system, the United States might take the initiative with other members of the Security Council to examine the feasibility of establishing a special peacekeeping fund. This initiative was included in the secretary of state's report to the president on reform of the UN system mentioned earlier, but the Reagan administration did little to advance the proposal in UN precincts. Nor has the United States made good on earlier pledges to make our resources available as the "eyes and ears" of UN peacekeeping operations.

Within the various bureaucracies and agencies of the U.S. government a greater sense of urgency and commitment to international peacekeeping is required. But the minds of senior officials will be focused on the subject and sustained over a period of time only if priority is accorded to peacekeeping by the president, his immediate staff, and leading members of Congress.

Recommendations

There is much more that the U.S. government can do to bolster the success of peacekeeping by third parties. First, security assistance funds allocated to peacekeeping should be increased. The U.S. security assistance program already makes provision for a special fund to finance UN peacekeeping operations in Lebanon and Cyprus. Congress might wish to examine the feasibility of (1) increasing allocations for the account to serve as seed money for the UN Special Peacekeeping Fund first proposed in 1978; (2) according higher priority than presently obtains for allocation of Economic Support Funds (ESF) to countries prepared to create, train, and reserve forces for international peacekeeping under UN auspices; and (3) setting aside earmarked funds for stockpiling advanced technology required for third-country peacekeeping purposes.

Resources available under the military assistance program might be utilized for peacekeeping as well. Section 506 of the Foreign Assistance Act makes provision for recourse to older Department of Defense stocks to meet U.S. national security needs. The secretary of defense might be invited by Congress to conduct a general survey of such stocks that might be available for third-party peacekeeping purposes, their general condition, and costs associated with their upgrading and maintenance.

Greater emphasis might be given to peacekeeping in our foreign military education and training programs. In view of the need to familiarize other nations willing to provide peacekeeping forces with applications of state-of-the-art technology, the U.S. government, within the framework of the Security Assistance Program, might wish to earmark a portion of its annual military education and training appropriations for such purposes. In addition, Washington may wish to explore with other allied or friendly governments the feasibility of pooling or coordinating newly organized training and education programs.

The United States should also seek to involve the Soviet Union in cooperative peacekeeping endeavors. The U.S. government should discuss peacekeeping operations as an integral part of its regularly scheduled meetings with the Soviet Union on Third World regional issues. Indeed, if the subject has particular merit, the assistant secretary of state for international organizations might be authorized to meet with his Soviet counterpart to develop an action agenda aimed at planning for joint airlift of peacekeeping contingents from other nations, equipment stockpiling, and cost sharing.

The private sector should also be invited to play an active role. Congress, together with appropriate executive branch agencies, should encourage the formation of private sector committees to examine the contributions that nongovernmental groups can make to strengthen international peacekeeping operations. The U.S. Chamber of Commerce, various trade associations, and retired officers' groups could well become the fulcrum for these activities.

Within the U.S. government, the applications of technology for peacekeeping have attracted only marginal attention. Given the instabilities of the Third World, several agencies should be devoting resources to in-depth studies. The Arms Control and Disarmament Agency, the Department of State (particularly the offices of the under secretary for security assistance, science and technology, and the bureau of international organization affairs), the science adviser to the president, and the Department of Defense come most readily to mind. In the para-official community, the National Academy of Sciences, RAND, and similar organizations and institutions should be encouraged to conduct their own inquiries.

In the political-military realm there are very real cost-benefit trade-offs to be examined where military technology is concerned. For example, U.S. and other Western producers might wish to weigh the advantages of stockpiling low-cost items such as night-vision devices, crowd-control equipment, and protective shields. When tactical surveillance equipment has reached advanced stages of evolution, coproduction and training arrangements with other producer countries might be assessed.

Finally, all of these efforts would have more chance of success if congressional sentiment could be marshalled for enhancement of peacekeeping efforts on the part of international organizations. Clearly, there exists today a predisposition to use U.S. forces to deal with Third World crisis situations. Congress might begin to explore the multilateral peacekeeping alternative by inviting executive branch representatives to testify on U.S. government plans and efforts to bolster the capacity of international organizations to assume broader peacekeeping roles. Congress might also develop legislation creating an equipment "set aside" within the security assistance program for peacekeeping purposes. And the War Powers Resolution might be amended to require that the president report to Congress on rescue, humanitarian, and peacekeeping operations by U.S. forces. In each of these critical areas, review and intervention by legislative leaders is essential if the United States is to play a significant role in placing technology at the service of conflict resolution within the strife-ridden community of nations.

8

SAFETY, SECURITY, AND CONTROL OF NUCLEAR WEAPONS

BY GERALD W. JOHNSON

From the beginning of the U.S. nuclear weapon program, the safety, security, and control of nuclear weapons has been of paramount concern to those responsible for their development, logistics, and operational deployments. As the size and diversity of the nuclear stockpile increased, and as deployments of the weapons became more widespread, the nature of problems associated with these activities became more complex. Yet problems of nuclear safety and security have been handled with remarkable success.

There has not been a single accidental or unauthorized nuclear explosion or seizure of a nuclear device in the more than forty years that such weapons have been in the U.S. inventory. Although a few accidents have resulted in the dispersal of nuclear materials over small areas, in every case contaminants were quickly reduced to safe levels. Nor have there been public reports of an accidental firing or unauthorized possession of a nuclear explosive anywhere else in the world since production of these weapons began. This record has been achieved despite the fact that tens of thousands of weapons have been deployed by at least six states—United States, Soviet Union, United Kingdom, France, China, and India—and perhaps as many as four others.

Despite this virtually perfect record, the U.S. Departments of Defense and Energy are continuing intensive efforts to improve the safety and security of nuclear weapons. In recent years, these efforts have concentrated on threats involving the possible seizure and attempted use of nuclear devices by terrorist groups. Such threats, of course, are no different than any other unauthorized seizure of a nuclear weapon, and so much of present work concentrates on the development of administrative arrangements to prevent the disclosure of information about the logistics of nuclear weapons and materials that could assist a terrorist group seeking to seize a weapon. The research and development of technical means to thwart terrorists is also continuing, however, concentrating on the design of "Active Protection Systems," devices that could be incorporated into weapons to cause automatic disablement if unauthorized attempts were made to circumvent the safety and security devices already built into the weapon's firing systems.

In this chapter, the types of safety and security problems involved in the prevention of accidental or unauthorized nuclear detonations are described, followed by a brief summary of the history of the development of safety and security devices. The consequences of those few instances in which nuclear devices were involved in accidents are then discussed. And actions being taken now within the U.S. government to further improve the safety and security of nuclear weapons are outlined, particularly those actions in regard to threats that could be posed by terrorist organizations.

Although this chapter concentrates primarily on actions taken by the United States to ensure the safety and security of nuclear weapons, these problems clearly must be confronted by each of the other nuclear powers as well. The chapter concludes, therefore, with a discussion of the advantages and disadvantages of initiating a dialogue with other nuclear weapon states to share information on threats to the safety and security of nuclear weapons.

Safety

To discuss safety, it is necessary first to identify a nuclear weapon's key components. These components must be designed to minimize the chance of an explosion at a time or place other than desired and authorized. All existing U.S. nuclear weapons consist of a chemical explosive component and a nuclear fuel component. To produce a full-scale nuclear explosion, the chemical explosive must first be detonated in a precisely controlled fashion.

Atomic bombs of two fundamentally different designs were used for the attacks against Japan in August 1945. The Hiroshima weapon was the "gun-type"; the one dropped on Nagasaki was the "implosion" design. (The gun-type weapon had not been tested previously, while the implosion-type had been fired successfully at full yield in New Mexico a few weeks earlier.)

To accomplish a nuclear explosion using the gun concept, two subcritical masses of highly enriched uranium (U–235) are placed at opposite ends of a gun barrel with a relatively large diameter (about eight inches). A uranium slug is loaded at the breech end, along with an explosive charge, which is used to propel the slug at high velocity into a mass of uranium at the opposite end of the barrel. Because there is insufficient U–235 in either the target uranium or the slug to initiate a nuclear chain reaction by itself, the system cannot produce an explosion until the slug and its propellant are both placed in the barrel. Once that occurs, there is the possibility that an accident could cause the two components to join and initiate the chain reaction. In recognition of this possibility, only a few days before launching the attack on Hiroshima it was decided as a

138

matter of prudence to complete the assembly of the bomb in the aircraft bomb bay itself and only after the plane had become airborne. Otherwise, it was reasoned, there would be a finite chance of a very large nuclear explosion in the event of a crash on take-off, which could have caused extensive damage to the critical U.S. base on Tinian.[1]

The Nagasaki weapon employed plutonium, rather than U–235, as its nuclear explosive. The neutronic properties of plutonium are such that the use of the gun concept to achieve criticality in the Nagasaki bomb was judged to be impractical. Accordingly, to achieve sufficiently high efficiency in the nuclear reaction, it was necessary to develop a design in which criticality could be realized much faster than for U–235. After much effort, the necessary rapid assembly was demonstrated successfully, using a spherical charge of conventional explosives and igniting it simultaneously over its entire surface. This was accomplished by distributing detonators over the surface of the conventional explosive sphere and firing them electrically to produce the desired uniform ignition. Through the use of explosive lenses, the detonation wave was made to converge symmetrically toward the center of the sphere. The plutonium fuel located at the center was thus compressed rapidly to supercriticality and, with the injection of a burst of neutrons at the proper time, produced a nuclear explosion.

Because the nuclear fuel of such an implosion weapon is buried at the center of a sphere of high explosives, the fuel cannot be removed readily to make the weapon safe from a nuclear explosion in the event that the conventional explosive is detonated accidentally. As a consequence, the bomb used against Nagasaki already contained its plutonium core when the aircraft took off. Moreover, the high explosive had all of its detonators installed. Thus, if there had been a crash on take-off, there might have been an explosion large enough to cause damage to the base.

Had the high explosives used in these first atomic bombs been set off accidentally by exposure to fire or by impact in a crash, there would have been a number of possible consequences: First, the nuclear components could have been dispersed over an area up to perhaps a square mile, with concentrations of radioactive materials at levels requiring clean-up activities. Second, a small nuclear reaction could have occurred that would have exacerbated the problem of radioactive contamination. And third, there was a remote possibility, particularly with implosion weapons, of a substantial nuclear explosion, which of course would have created special problems if it had occurred in a populated area. For a full-scale nuclear

---

[1]Two, somewhat different, accounts of the handling of the Hiroshima bomb are given by David Hawkins, Edith C. Truslow, and Ralph Carlisle Smith, *Project Y: The Los Alamos Story* 40th Anniversary edition (San Francisco: Tomash Publishers, 1983), 253–4; and B.J. O'Keefe, *Nuclear Hostages* (Boston: Houghton Mifflin Co., 1983), 95–96.

explosion to have occurred, however, a series of conditions would have to have been met.

Just after the war, procedures were developed to remove this danger: implosion weapons were provided with separable, critical nuclear components that were inserted in the bomb—manually at first, and later mechanically—on command from the pilot of the aircraft, a process termed "in-flight insertion." It was impossible for the warhead under any circumstances to produce a nuclear explosion until the critical component had been properly put in place. The Atomic Energy Commission (later replaced by a component of the Department of Energy), moreover, retained physical custody of the separable component until its transfer to the military unit was authorized by the president.

By the late 1950s, the possibility of an accidental reaction also had been eliminated. And although the danger persists that nuclear materials might be dispersed as the result of an accident, the probability of even this has been greatly reduced. Still, important problems remain.

The first problem is posed by the lingering danger of an accidental high-explosive detonation. Technicians have, of course, accumulated an enormous amount of experience with the handling and use of chemical explosives, and technological innovation has made it possible to select explosives with characteristics suitable for various uses and conditions. As a general safety guideline, for any specific application, it is important to select or develop the most insensitive explosive consistent with the requirements of its designated function. These insensitive explosives require auxiliary firing systems to set them off. In commercial practice, such auxiliary elements are kept physically separated from the main charge until it is desired to fire the primary explosive. In modern nuclear weapons, however, all components are organic to the warhead. Thus even if a warhead could be made almost completely insensitive to external stimuli, it is still possible that nuclear material could be dispersed as the result of an accidental chemical explosion during the operational handling and deployment of a nuclear weapon. In the event of such an explosion radioactive nuclear materials—either U–235 or Plutonium—could be distributed across an area as large as a square mile, depending upon the nature of the explosion and prevailing weather conditions. (Even in the worst case, however, U.S. government agencies would be capable of reducing the concentrations of radioactive materials to levels well below established health limits.)

A second danger arises from the possibility of an accidental low-yield nuclear explosion. In the mid–1950s, the weapon design laboratories became concerned that if a weapon's conventional explosive were set off accidentally, a low-yield nuclear explosion, releasing perhaps tens of tons of nuclear energy, might take place. To address this concern, the laboratories initiated a design and test effort to determine the magnitudes of the yields that might be expected from such accidents and to develop new weapons that would produce a nuclear yield of no more than four pounds of high explosive equivalent for any conceivable accidental detonation of its conventional explosive.

Field tests were conducted at the Nevada Test Site beginning in November 1955 to ensure that in any accident, even if the conventional explosive were detonated, there would be no nuclear reaction.[2] The aim was to design warheads which would be "one point safe"—that is, a detonation initiated by a single detonator or by any other stimulus at any single point would not result in release of nuclear energy. (The operational definition for one-point safety is that the probability that the nuclear yield will be in excess of four pounds of TNT is less than one part in a million).

One byproduct of this testing was the discovery that if the high explosive detonated or burned, the plutonium or uranium fuel would be fragmented and could burn to produce finely divided radioactive oxide contaminants. The resulting levels and areas contaminated by plutonium were carefully documented and effective clean-up procedures developed.

During the U.S.-Soviet test moratorium (1958–61), work to provide for "one point safety" was continued by using subcritical quantities of fissionable material in an explosive assembly and measuring the neutron multiplication. In some of the tests slight releases of nuclear energy occurred, but the largest such energy release amounted to four-tenths of a pound of high explosive equivalent. These experiments served to identify critical safety issues and to resolve some of them, but a number of questions remained that could be dealt with only after the moratorium ended in 1961.[3] Today, no weapon can be introduced into the stockpile unless the possible nuclear energy release in an accident is very small (less than four pounds), and all nuclear weapons currently in the U.S. stockpile meet this criterion. Thus the probability of a nuclear explosion, even one of very low yield, following any conceivable accident involving U.S. weapons is near zero.

A third important area of concern is associated with the possibility of an accidental full-scale nuclear detonation. There are, of course, mechanisms in every nuclear weapon that, if properly activated, would produce the full, designed nuclear yield. The firing set—or the arming, fusing, and firing sequence—when initiated, will produce the necessary signal to cause simultaneous detonation of the chemical explosive and a symmetrical implosion of the nuclear material, resulting in a nuclear explosion. Consequently, before any new warhead is declared operational,

---

[2]Announced United States Nuclear Tests, July 1945 through December 1986, Department of Energy publication NVO–209 (Rev. 7, January 1987).
[3]Robert N. Thorn and Donald R. Westervelt, *Report LA–10902–MS, UC–2* (Los Alamos National Laboratory, February 1987).

a detailed analysis of the weapon system is made by a joint Department of Defense and Department of Energy safety group to ensure that safety criteria have been met.

Security

Questions of the security of fissionable nuclear material and nuclear weapons are more difficult to discuss because of stringent controls on information that might compromise either U.S. weapon designs or handling and storage procedures. The problem of the security and custodial control of nuclear weapons drew increasing attention in the late 1950s and early 1960s as the number of deployed systems increased and as deployments were extended beyond the continental United States.

Special concerns arose in the late 1950s when the armed forces of nations allied with the United States in the North Atlantic Treaty Organization (NATO) were assigned missions that included the delivery of U.S. nuclear weapons. It became possible, for example, that the armed U.S. custodians might be overpowered and that the European pilot of a bomb-loaded aircraft maintained on quick reaction alert (QRA) could take off without authority and deliver the bomb to any target of choice. Another concern was that an unauthorized individual or group would somehow acquire a weapon and use it to blackmail the government. Recently, similar fears have been expressed regarding the possible acquisition of a weapon by a terrorist group.

To address these concerns, the Armed Forces Special Weapons Project (now the Defense Nuclear Agency) initiated safety studies in 1957 that led to the development of Environmental Sensing Devices (ESDs). These devices, which have since been introduced into several U.S. nuclear weapons, factor in weapons' associated delivery systems to ensure that even in the operational mode and in the normal firing sequence a nuclear explosion cannot occur until the weapon has also sensed appropriate environmental conditions. The basic function of the ESD is to preclude the accidental or premature detonation of the nuclear explosion under conditions that would endanger launching systems, friendly forces, or major facilities or populations. To accomplish this, switches are added to the firing circuit which remain either open or closed depending upon their design and the nature of the environment to which they are intended to react. Nuclear bombs attached to an operational bomber, for example, incorporate a temperature sensing switch in the firing circuit that opens only when the weapon reaches a sufficiently high temperature; otherwise the switch prevents detonation of the bomb, even in the event of a crash. Switches in other types of weapons are activated only when high acceleration or spin rates are experienced. Warheads delivered by tactical rockets or artillery projectiles incorporate ESDs that do not allow warheads to explode until they are beyond ranges that would put the firing crew at risk.

Another class of problems emerged as U.S. nuclear weapons began to be deployed to Western Europe for employment by allied armed forces in the event of war. Such deployments were authorized in 1958 by an amendment to the Atomic Energy Act of 1946, with the proviso that any nuclear weapons so deployed must remain in the possession of U.S. custodians until their release by the president for use in war. The Congressional Joint Committee on Atomic Energy, noting that U.S. nuclear weapons would be attached to NATO aircraft and missiles, pondered the security of such arrangements to ensure the "custody" responsibilities by the United States under the Atomic Energy Act. Of particular concern were the warheads slated for assignment to fighter aircraft (the B7 bomb) and the Thor and Jupiter intermediate-range ballistic missiles (the W49 warhead). The Los Alamos, Livermore, and Sandia National Laboratories explored possible technical approaches on their own, including the construction of breadboard components.

Because the Atomic Demolition Munitions were intended to be emplaced manually in prepared holes, a suitable ESD could not be devised; they were instead provided with three–digit combination locks that physically blocked access to key arming elements. By mid-1960, researchers at Sandia had identified a number of approaches to a remotely operated combination lock for these warheads. This earlier effort for atomic demolition munitions was extended directly to the more general problem as well.

The transfer of certain restricted data and nonnuclear components to selected NATO allies was authorized by Congress in 1958. After bilateral agreements had been negotiated, certain NATO forces were trained and delivery systems equipped for nuclear warheads to be provided by the United States under appropriate custody and control procedures. Shortly thereafter, additional nuclear warheads began to be moved to Europe. The early deployments involved bombs with separable components; the latter, of course, were to remain physically in American custody until released by order of the president of the United States.

Permissive Action Links

A new issue emerged in 1960 with the decision by the Supreme Allied Commander in Europe to place some of his forces on quick reaction alert to reduce the chance of their destruction on the ground in a surprise attack. This meant that fully assembled bombs and warheads would be loaded on planes and missiles, ready to be launched in minutes. These weapons were of the new "sealed pit" design and were ready to fire once properly armed.

Concurrently, serious studies of the general problem of custody and control were initiated and, late in 1960, the Joint Atomic Energy Committee and its staff made a tour of NATO to review the storage, logistics, and operating arrangements. These studies produced a number of suggestions to deal with the identified problems, which were forwarded to the president in early 1961. The central recommendation was to design coded electromechanical devices that would be incorporated in the warheads themselves and would prevent the unauthorized use of the weapons.[4]

A technical solution to the issues raised by the Joint Committee on Atomic Energy was jointly worked out by the Sandia and Los Alamos laboratories. The concept was to embed a mechanical or electromechanical coded switch in the warhead in a location such that it could not be bypassed readily. To foil any attempt to bypass the device, the switch's appearance and markings were disguised to make its function unclear unless the weapon's manual were also available. The code was also made sufficiently complex that to decode it would require a minimum of at least several hours. These new switches were first termed "permissive links," but later became known as "permissive action links" (PALs). The primary function of the PALs was to prevent unauthorized use of nuclear weapons; and although they were not required for increased safety, this was nevertheless a tangential benefit of the devices.

The switch created a break in the arming or firing circuits that could only be closed by the insertion of the proper code. To retain control, the code was made known only to U.S. command at an appropriate operational level, with secure communications between the authorized user of the weapon and the U.S. military officer who had the code. The codes were changed from time to time and, as improved switches were developed, warheads were retrofitted with them. As a further safeguard, the PAL code system was split into two parts so that no one person had access to the full code until the weapons were released for use. PAL teams in Europe still use this system today for both four- and six-digit codes.

Political and technical activity to implement the PAL concept increased in 1962. The science adviser to the president, Jerome Wiesner, provided a set of recommendations that were incorporated in a National Security Action Memorandum (NSAM 160) issued on June 6, 1962. NSAM 160 ordered that all U.S. tactical nuclear warheads assigned to NATO be equipped with a PAL, and within three months all deployed Jupiter missiles were equipped with the first PALs. (Ten years later, this order was extended to the control of selected weapons in the Pacific.) The memorandum also established a continuing research program to improve the coded switches and to ensure their integration into new tactical warhead designs as the latter emerged from development programs.

The immediate task in 1962 was to produce and install on a crash basis an electromechanical four-digit lock on bombs for QRA aircraft in

---

[4]"Nuclear Safety Devices and Laymen on the Atomic Energy Commission," *Congressional Record*, vol. 108, no. 116 (July 10, 1962).

Europe and on warheads for the Thor and Jupiter missiles deployed in the United Kingdom, Italy, and Turkey. At the urging of Secretary of Defense Robert S. McNamara switches were installed by the fall of 1962.

Development continued in parallel: PALs for missiles were designated category A and those for bombs, category B. The new units were retrofitted to the Jupiter IRBM (with the W49 warhead), to gravity bombs (with the B7 and B28), and to tactical missiles—Mace (W28), Pershing (W50), and Sergeant (W52).

Later, category B PALs were improved and a new category C device was introduced. The former required fewer wires, permitting its installation in the cockpit and allowing the pilot to unlock the PAL when he received the proper authorization. Other features of the category B device included an ability to recode and to check the operational code without unlocking the PAL, decreased operational time, and a code-controlled lock operation. These units were installed in a number of gravity bombs—the B28, B43, B57, and B61-0.

In 1965, steps were taken to upgrade the category B PAL to incorporate a "limited try" feature that would preclude an exhaustive search to decode the switch. The feature disabled the the nuclear weapon after a certain number of erroneous codes had been attempted. This improved switch was used only in the Walleye guided glide bomb (W72), but in 1973 a six–digit, limited-try category C PAL was introduced for the Lance missile (W70-0).

PAL devices can be either combination locks or coded switches. The combination locks provide a mechanical barrier that, through a variety of means, inhibits the firing of the weapon. Coded switches maintain the warhead in an electrically disabled condition until the proper code is inserted.

All PAL units are now designed to permit unlocking, recoding, and verification operations. Combination locks do not require any external components to operate, while coded switches do. The most recent PALs, category D, are based on twenty-five years of engineering effort and have been tailored to new systems like the ground-launched cruise missiles now being deployed in Europe. These units incorporate "limited try" capabilities, multiple code capabilities, and unlock populations of at least one million. (The applications of the various PALs now in use or being deployed are listed in table 1.)

Incidents and Accidents

The combination of basic warhead design safety features, PALs, and associated administrative procedures have successfully protected codes and controlled access to U.S. nuclear weapons in Europe with the result that there has been no unauthorized use and no accidental nuclear explosion. There have been a number of incidents over the years, however,

TABLE 1
CROSS-INDEX: WARHEAD/SERVICE/PAL USE CONTROL CATEGORY

Warhead	Service	PAL Category
B28–0.1	Air Force	D
W31	Army	D
B43–2	Air Force	B
W50–1	Army	A
B57–2	Air Force	B
B61–0	Air Force	B
B61–2,5,7	Air Force	D
B61–3,4,6,8	Air Force	F
W70–1,2,3	Army	D
W79	Army	D
W80–0	Navy	D
W80–1	Air Force	D
W82	Army	D
B83	Air Force	D
W82	Army	D
W84	Air Force	F
W85	Army	F

which have raised certain concerns and prompted a number of technical and operational responses. These incidents included an accidental detonation of a weapon's high-explosive component, accompanied by the limited dispersal of its fissionable material, requiring costly clean-up operations.

A summary of all of the nuclear accidents between 1950 and April 1981 prepared by the Defense Department and the Department of Energy identified thirty-two events, none of which resulted in the release of nuclear energy and only two of which resulted in the local dispersal of radioactive materials. Of these events, twenty-seven were associated with aircraft—including crashes, fires on the ground, and the jettisoning of bombs in emergencies—and only two involved ballistic missiles. (See Appendix.)

When the Soviet Union began to deploy ICBMs in the 1950s, it was widely believed that the short flight time of the weapons placed the U.S. Strategic Air Command (SAC) bombers at risk of surprise attack. To meet this threat, SAC instituted a new alert system, in which some fraction of the U.S. bomber force was maintained continuously in the air and a second fraction was maintained on alert near the runway with the crews nearly ready to take off if ordered; aircraft on the runway—so-called strip alert—and on airborne alert were fully loaded with nuclear weapons ready for operational missions. Because this new SAC posture increased the chances of an accident, laboratories began to design nuclear weapons that would not explode as the result of crashes, fires, or similar accidents.

Although, as noted earlier, field tests had begun as early as November 1955 to design "one point safe" warheads, additional research was needed to assess the possible level of danger from accidental spreading of fissionable materials. Experiments using explosives were conducted in Nevada during Project 56 (1955–56), Project 57 (1957), and Roller Coaster (1963) to determine the distribution of the fissionable materials in the event of such an accident, the avenues for uptake of these nuclides in human beings, their long-term effects, and methods to decontaminate areas to safe levels in case of an accidental dispersal.

Drawing on the results of these tests and other experience, AEC and DOE have established rigid criteria to ensure that any threat to human life from any release of plutonium is minimized. In any cleanup of plutonium contamination, the levels would be reduced far below toxic levels.

A number of accidents involving airborne alert aircraft have occurred—including one of particular interest on January 17, 1966, in Spain.[5] A B–52 carrying four thermonuclear weapons on airborne alert had rendezvoused at 30,000 feet with a KC–135 tanker over Palomares, Spain, during a routine refueling operation. The planes caught fire and disintegrated in the air. The four bombs fell to the earth—three striking land and the fourth sinking in the sea.

The weapons on this particular flight were fitted with parachutes designed to permit them to settle to the surface of the earth before detonating, if that were desired for tactical reasons. One of the bombs that struck the land and the one that fell in the ocean were recovered essentially undamaged because their parachutes deployed properly. The parachutes on the other two bombs did not deploy, however, and the bombs broke open and released the plutonium and U–235 in finely divided form. All of the explosive in one of the bombs detonated while in the other there was only a partial detonation leaving some of the unburned high explosive scattered about. In neither instance was there any release of nuclear energy. The resulting contamination was similar in amount and dispersal to what had been observed in the earlier experimental Nevada tests.

Teams of experts, some of whom had participated in the Nevada tests, had been organized and equipped to respond to just such an emergency as the Palomares incident. The teams arrived in Palomares the day following

---

[5]Tad Szulc, *The Bombs of Palomares* (New York: Viking, 1967); Flora Lewis, *One of Our H-Bombs is Missing* (New York: McGraw-Hill, 1967).

the incident and went to work, first locating the bombs on land, which required a search of about four hours. The intact bomb was made completely safe by an Explosive Ordnance Disposal team and was immediately taken under protective custody by the U.S. Air Force. The other two bombs were contaminated and had to be handled by the team of experts to ensure their safe removal.

The contaminated area, amounting to about 604 acres, was carefully mapped and, in three weeks of cleanup, 7,500 tons of the most highly contaminated material were loaded in fifty fifty-gallon drums, sealed, and shipped to South Carolina for storage at the Savannah River site of the AEC. Less contaminated areas were simply plowed under—an effective technique first demonstrated in Nevada—to reduce concentrations of radioactive material far below accepted standards. To locate and recover the bomb in the ocean proved more difficult; but after three months the bomb was recovered intact.

A similar incident occurred near Thule, Greenland, on January 21, 1968, when a B–52 crashed and burned while approaching a U.S. Air Force base. There were four nuclear weapons aboard the B–52, all of which were destroyed in the fire, distributing radioactive debris on nearby sea ice. To reduce the contamination to the very low levels required by U.S. government standards, some 237,000 cubic feet of contaminated ice, snow, and water, together with crash debris, were shipped to an approved storage site in the United States. Because of concerns aroused by these two incidents as well as budgetary considerations, airborne alerts of nuclear-armed bombers were terminated in 1968.

Incidents involving nuclear-armed missiles also have occurred. On September 19, 1980, near Damascus, Arkansas, a Titan II ICBM with a warhead mated to it was accidentally damaged in the course of a maintenance procedure, causing its liquid fuel to leak. It was not possible to seal the leak, so all personnel left the silo. A few hours later, an explosion removed the silo door and lobbed the warhead out to the surface. There was no release of radioactivity, however, and it was reported that the warhead was recovered essentially intact.

A similar incident took place on October 3, 1986, when a Soviet Yankee-class ballistic missile submarine on a routine patrol 1,300 miles off the U.S. Atlantic coast experienced a fire and explosion in one of its missile tubes.[6] The cause of the explosion has not been reported, but it was sufficiently powerful to rip open the cover of the missile tube and to cause such internal damage to the submarine that it sank four days later.

---

[6]The concept is described in a now declassified letter, dated January 5, 1961, addressed by Harold Agnew, a physicist at Los Alamos, to Major General Alfred Starbird, director of the Division of Military Applications of the AEC.

The SS-N-6 missile deployed on *Yankee*-class submarines is liquid-fueled. What became of the warhead deployed on the missile that exploded or the other fifteen warheads on the submarine has not been reported, but it is clear that if any nuclear energy was released, it was very small. Sampling of the air and water in the area by U.S. forces did not reveal any radioactivity above normal background levels, suggesting that the high explosive in the warhead may not have exploded and that there was no dispersal of plutonium. Even if some plutonium had been deposited on the hull of the submarine, the washing action of the heavy seas prevalent at the time would have rapidly diluted it, and it would not have been detected in later monitoring. In any case, it is clear that even under the severe conditions of the accident there was no nuclear explosion.

The Future

During the first forty years of the nuclear era, there have been no reported or observed accidental nuclear explosions or unauthorized firings. Since 1945, all nuclear explosions have been conducted by governmental agencies for the purposes of the further development of the weapons themselves, for measurement of nuclear effects, and for constructive and scientific applications. There have been a number of incidents in which deployed weapons were exposed to extreme environmental conditions resulting in the explosion of the high-explosive component and in the distribution of fissionable material over limited areas. For those areas in which the contamination levels exceeded government guidelines, however, it was possible in each case to carry out clean-up procedures that reduced contamination well below safety specifications.

All of this is favorable history. Yet the rise of state terrorism in recent years suggests a need to review the possible nuclear dangers posed by this new threat and consider what additional measures should be taken to counter it. The priorities must be, first, to continue efforts to reduce opportunities for unauthorized individuals to seize a weapon to a minimum consistent with operational requirements, and, second, if, in spite of all precautions, a weapon somehow were acquired, to complicate its successful firing to the maximum extent possible.

To deal with the first problem, there should be a continuing review of the administrative and security procedures now in place for the manufacture, storage, reworking, and transportation of nuclear weapons in all phases of the logistical chain. The establishment of a knowledgeable, high-level group within the Department of Defense charged with this responsibility could provide the leadership necessary to ensure continuing attention to this critical matter. Through the identification of possible weak links in the logistical chain, and with detailed knowledge of those threats that are identified, the group could take the necessary steps to foreclose any opportunities for a determined team to acquire a weapon. An important element in this effort would be the use of realistic techniques in gaming possible terrorist operations. In principle, the kinds of steps to be emphasized would involve limiting public disclosure of patterns and schedules of operations, changing operations—or at least their appearance—in unpredictable ways, and shaping security procedures to complicate and deter threats to critical links. An important, perhaps the most important, requirement of an effective counter-terrorist strategy is hard, current intelligence on terrorist threats. The Reagan administration gave considerable attention to this problem.

With regard to the problem of preventing a detonation in the event of a security failure, we have described the introduction of PALs and have observed that current and planned designs greatly complicate any attempts to wire around the devices. Although it remains possible that a competent team might conceivably succeed in preparing a warhead for firing, new technical approaches are being explored at the national laboratories to reduce even this remote possibility. With the new "Active Protection Systems," for example, if an attempt were made to circumvent the safety device, a disabling action would automatically damage the warhead so severely that it could not explode even if the proper firing sequence were followed. In fact, restoring such disabled warheads to an operational status would require that they be returned to a production facility, dismantled, and repaired.

Substantial resources will be required to sustain present levels of safety and security. Those charged with planning and managing nuclear safety and security must be provided the means to locate weak links at any point in time and to take the steps necessary to minimize the danger of the unauthorized acquisition and firing of any nuclear warhead under the control of the United States.

This discussion has been limited to U.S. practice and experience with respect to the safety and security of nuclear weapons. But it is clear that there is a common interest on the part of other nuclear powers to achieve the same objectives. There have been some joint efforts with the United Kingdom, and the time may have come to explore with other powers, including China and the Soviet Union, common interests and concerns and to work out common approaches for dealing with them. Some kinds of cooperative efforts, especially to meet terrorist threats, could be identified and implemented. What is needed is the establishment of a process to encourage continuing exchanges of information and cooperative efforts drawing on the past experiences of others. By such procedures, the successes achieved in the past will best be ensured for the future.

---

[7]Bernard Gwertzman, "Soviet Submarine," *New York Times*, October 6, 1986, A–1; and George C. Wilson and R. Jeffrey Smith, "Soviet Sub Sinks," *Washington Post*, October 7, 1986, A–1.

9

REDUCING THE RISK OF NUCLEAR WAR WITH PERMISSIVE ACTION LINKS

BY DAN CALDWELL AND PETER D. ZIMMERMAN

During the first several months of 1967, the military commander of the Chinese province of Xinjiang, General Wang En-Mao, threatened to take over the Chinese nuclear testing facility (and presumably the nuclear weapons stored there) at Lop Nor in a dispute with the central government, then headed by Mao Zedong.[1] Eleven years later, in 1978, three men were arrested in a bizarre plan to steal and sell an American nuclear submarine, the USS *Trepang*.[2] In neither of these cases were the nuclear weapons protected against unauthorized use. These incidents illustrate some of the possible means, however remote, through which nuclear weapons belonging to any nation could be obtained by individuals not authorized to use them.

In the 1967 Chinese incident, the dispute between General Wang and Chairman Mao was resolved, and Wang's threat to take over Lop Nor came to nought. But the danger had been made clear: a military commander could gain independent control over nuclear weapons. In the 1978 case, the three would-be submarine thieves had planned to steal the sub and then to sell it to the Mafia. To create a diversion, the three men—one of whom was a navy veteran who had served on the *Trepang*—had planned to fire a nuclear-tipped torpedo at an eastern U.S. city. As implausible as this plan may have seemed, the FBI thought it serious enough to assign twenty agents to the investigation.

Since the early 1960s, the United States and several other nuclear weapons states have placed use-control devices on many of their nuclear weapons. The United States has installed use-control devices on all of its

---

[1] C. L. Sulzberger, "Foreign Affairs: The Nuclear Pawn," *New York Times*, February 5, 1967, E-8; Richard Hughes, "Mao Calls Truce with Rebel General in Bomb Province," *London Sunday Times*, July 2, 1967; and H. M. Jones, "China: Autonomous Wang," *Far East Economic Review*, December 28, 1967, 569–70.

[2] Charles R. Babcock, "Suspects in Plot May Have Thought Mafia Wanted Sub," *Washington Post*, October 6, 1978, A–12. We are indebted to Leonard Spector for calling this incident as well as the Wang En-Mao affair to our attention.

land-based nuclear weapons. The most common use-control devices are permissive action links (PALs). In their simplest form, permissive action links are mechanical or electronic combination locks placed on nuclear weapons; without the combination the weapons cannot be detonated. More advanced PALs also incorporate "limited try" features and protective skins around the permissive action link mechanism. Given the sophistication and reliability of modern PALs, it is virtually impossible that a deranged government official, military officer, terrorist, or any other unauthorized individual could detonate a weapon equipped with such a device.[3]

Unfortunately, not all nuclear weapon states have PALs installed on their nuclear weapons, and, in fact, not all U.S. nuclear weapons are equipped with PALs. The most numerous class of U.S. nuclear weapons without PALs are those deployed on U.S. Navy ships and submarines.

Threats of unauthorized nuclear detonations could be lessened world-wide if the United States adopted two policies: to share information on safety devices to prevent unauthorized detonation with other nuclear weapon states and to place permissive action links on those nuclear weapons that are not currently equipped with such devices, particularly nuclear weapons on naval vessels. This chapter assesses the desirability and feasibility of these two policies.

In this chapter the development of permissive action links by the United States and the design principles on which PALs are based are described; nuclear weapons that have PALs and those that do not are identified; and several policy alternatives are proposed for placing PALs on the nuclear weapons of other states and on U.S. naval weapons.

The Development of PALs by the United States

On the recommendation of the Department of Defense, in 1958 the U.S. Congress amended the Atomic Energy Act to allow for the deployment of nuclear weapons with U.S. allies in Europe. In addition, the Defense Department proposed that military personnel from NATO countries receive training concerning the arming and firing of these weapons.[4] Within four months, Thor and Jupiter intermediate-range ballistic missiles were being installed in the United Kingdom, Italy, and Turkey, and military personnel from these countries were being trained in the operation of nuclear weapons.

American officials were naturally concerned that the United States maintain control over these weapons, and several measures were adopted

---

[3]Peter Stein and Peter Feaver, *Assuring Control of Nuclear Weapons: The Evolution of Permissive Action Links,* CSIA Occasional Paper 2 (Lanham, Maryland: University Press of America for the Center for Science and International Affairs, Harvard University, 1987), 58.

[4]Joel Larus, *Nuclear Weapons Safety and the Common Defense* (Columbus, Ohio: Ohio State University Press, 1967), 80–81.

to ensure this control. First, nuclear weapons were to be strictly guarded. Second, the "two-key" system was adopted. According to this procedure, two keys were required to arm a nuclear weapon. One key was to be in the possession of a British, Italian, or Turkish officer and the other on the person of an American officer; both officers would have to agree for the weapon to be armed. Third, during the mid– to late 1950s, engineers at the Atomic Energy Commission laboratories developed "environmental sensing devices," which, according to nuclear weapons expert Donald Cotter, monitor "a number of different 'environments' that can be duplicated only in the flight of a bomb; close-to-zero-gravity accelerations [that is, free fall], changes in barometric pressure, and deceleration caused by deployment of a parachute to slow the bomb's descent. Timers are used in some cases to ensure that these environments occur in a proper sequence and time frame."[5] These technical measures restricted the physical circumstances under which a nuclear weapon could be exploded, but they did little to prevent wholly unauthorized employment of the weapons. A series of incidents was soon to demonstrate the need for greater security.

In June 1959 a U.S. congressional delegation inspected a number of bases in Europe where U.S. nuclear weapons were deployed. While visiting a Thor missile base at Feltwell, England, Representative Charles Porter found that the British missile control officer had possession of both his key and the second key that was supposed to be in the possession of an American officer. Congressman Porter noted that U.S. control had been essentially short-circuited by this violation of the regulations.

In 1960, the Supreme Allied Commander in Europe placed a number of his forces on "quick reaction alert," which resulted in NATO forces and not just U.S. forces having nuclear warheads mated to their delivery systems ready to be used. Members of the Joint Committee on Atomic Energy went to Europe in December 1960 to inspect the security of nuclear weapons at NATO bases. At several bases they found a quick reaction alert airplane loaded with armed nuclear weapons and a foreign pilot in the cockpit. According to one member on this trip, "We landed in West Germany and saw all of these German planes with German [Maltese] crosses and an eighteen-year-old American private with a rifle to guard U.S. nuclear weapons. I was not reassured."[6] Upon his return, one of the members of the delegation, Harold Agnew, wrote to the director of the military applications division of the AEC and noted the problems that the delegation had seen. Agnew recommended "that a coded arming device be

---

[5]Donald R. Cotter, "Peacetime Operations: Safety and Scope," in Ashton B. Carter, John D. Steinbruner, and Charles A. Zraket, eds., *Managing Nuclear Operations* (Washington, D.C.: The Brookings Institution, 1987), 47.

[6]Author's confidential interview.

provided in those NATO weapons which must have a ready [i.e., quick reaction alert] capability."[7]

Several people in and out of government had been thinking along similar lines. In 1958, Fred Iklé, a social scientist at the RAND Corporation, had suggested that a "combination lock" be installed on nuclear weapons to insure better control.[8] Others in the Department of Defense and the nuclear weapons labs also advocated the permissive action link concept, but there was opposition from some military officers to PALs. With support from the Joint Committee on Atomic Energy, civilians in the Department of Defense, and President John F. Kennedy himself, National Security Action Memorandum (NSAM) 160 was signed in June 1962. It mandated that all nuclear weapons deployed in Europe be equipped with PALs. Within a remarkably short period of time—four months—PALs were being installed in both NATO and U.S.-controlled nuclear weapons based in Europe.

NSAM 160 did not direct that PALs be installed on nuclear weapons deployed by either the Strategic Air Command (SAC) or the U.S. Navy. These exemptions were made because PALs were designed initially to ensure U.S. control over nuclear weapons deployed outside of the United States, and both navy and SAC weapons were not believed subject to hostile take-overs. Over time, however, permissive action links or functionally equivalent use-control devices have been placed on all SAC weapons.[9] Only the U.S. Navy remains exempt.

Non-U.S. Use-Control Devices

Following the installation of PALs on some U.S. nuclear weapons, other countries developed their own procedures and devices to ensure the safety and security of their nuclear weapons; some of these were similar to American procedures and technology and some were different.[10] The available evidence indicates that in addition to the United States, the Soviet Union, Great Britain, and France have permissive action links or similar use-control devices on at least some of their nuclear weapons. It does not appear that China has PALs on its weapons, a deficiency that poses both domestic and international dangers to China.

The 1967 incident involving General Wang illustrates one type of domestic threat when PALs are not placed on nuclear weapons. A subsequent

---

[7]Letter from Harold M. Agnew to Major General A.D. Starbird, January 5, 1961, declassified by the U.S. Department of Energy on August 28, 1985.

[8]Fred Iklé with Gerald J. Aronson and Albert Madansky, "On the Risk of an Accidental or Unauthorized Detonation," RM–2251 (Santa Monica, California: RAND Corporation, 1958).

[9]Stein and Feaver, *Assuring Control of Nuclear Weapons,* note 3, 65.

[10]For a more detailed description of non-U.S. use control devices, see Dan Caldwell, "Permissive Action Links: A Description and Proposal," *Survival,* vol. 29, no. 3 (May–June 1987): 224–38.

incident in China underscored the need for assured central governmental control over nuclear weapons. In 1971, Lin Biao, defense minister at the time, attempted to overthrow Mao. If Lin had gained control over China's nuclear weapons during this insurrection, what might have happened? His, or any other dissident faction, could have conceivably blackmailed the central government by gaining control over some of China's nuclear weapons.

This situation is not unique to China. In 1960, following an unsuccessful *coup d'état* by a group of French generals who opposed his policies in Algeria, Charles de Gaulle ordered the detonation of one of France's nuclear test weapons so that it could not fall into the hands of other would-be revolutionaries.[11]

The absence of permissive action links on Chinese nuclear weapons could have had serious international consequences. In March 1969, the Soviet Union and China almost went to war as a result of a military confrontation in the Ussuri/Amur River area. According to a number of reports, during this crisis the Soviet Union considered attacking Chinese nuclear facilities preemptively. What if a similar incident were to occur today? China now has many more nuclear weapons than in 1969, increasing Soviet apprehensions. Given the current Chinese nuclear arsenal and the traditional Russian fear of China, these tensions would be increased if the Soviets thought that the Chinese central government did not have firm control over its nuclear stockpile. Taken to the extreme, Soviet anxieties over the possible loss of central control over China's nuclear weapons could lead to a Soviet preemptive attack. If Chinese nuclear weapons were equipped with permissive action links and if Soviet leaders knew this, these apprehensions might not be so acute, and the danger of nuclear conflict would decrease.

There is another important group of states that have either developed nuclear weapons secretly or are close to developing a nuclear weapons capability. This group includes Israel (which most experts believe possesses the capability to assemble quickly up to several hundred nuclear weapons), India, South Africa, and Pakistan. Given the transfer of nuclear know-how and technology that has taken place from the United States to Israel, it is possible that Israel already has the capability to develop PALs. It is unlikely, however, that India, South Africa, or Pakistan have permissive action links on their weapons (if indeed they have weapons).

As noted with regard to China, there are significant national security reasons for these countries to place PALs on their nuclear weapons. First, PALs ensure that a centralized government has control over its nuclear

---

[11]Donald G. Brennan, "The Risks of Spreading Weapons: A Historical Case," *Arms Control and Disarmament,* vol. 1 (1968): 86.

weapons. Given existing factional strife in the three countries under consideration, this alone would be reason enough to consider seriously the emplacement of PALs on nuclear weapons. Second, permissive action links reduce the danger of unauthorized nuclear detonations. Third, PALs can reduce the possibility that crises might result in nuclear war. Permissive action links are not complicated to build, but political will and determination are needed to subdue the bureaucratic resistance to placing them on all nuclear weapons. In fact, it is unlikely that the United States would have installed PALs on its nuclear weapons without the support of President Kennedy in the early 1960s.

U.S. Permissive Action Links

Roughly speaking, a use-control device for an atomic bomb is nothing but a strongbox outfitted with a good lock and containing a nuclear weapon. In fact, early security systems for U.S. nuclear weapons apparently consisted of just that, sealed containers with tamper-resistant combination locks.

"Tamper *resistant*" is the best result a designer can hope to achieve; it is clearly impossible to construct a locking system that is absolutely tamper *proof*. Even so, in designing PALs, trade-offs must be made among several variables, including complexity of operation, cost, reliability, and security. In general, a designer must be content with specifying the length of time that a would-be unauthorized user or potential thief would be delayed rather than seeking to guarantee the absolute security of the protected nuclear device.

For the purpose of discussing the principles of PAL design, a nuclear weapon can be thought of, schematically, as an onion with many layers, or as a Russian doll *(matryoshka)* with progressively smaller dolls nested within the larger dolls. A PAL designer can locate a use-control device outside the case of the weapon, but placed within the aircraft that would be used to deliver the bomb, inside the weapon's firing circuitry, inside the explosive charge that triggers the nuclear explosion, or even within the special nuclear material itself. The deeper the PAL is embedded in the weapon, the earlier it must be worked into the design and the more carefully and extensively it must be tested. It is, in fact, the case that safer weapons might require at least some testing of their nuclear components.

Locking mechanisms of many kinds can be used in PAL designs, although the most common systems use electromechanical devices, often with alphabetically coded switches. The simplest lock uses a securely designed key—often two different keys of different designs that must be turned simultaneously. Keys alone, however, do not, in fact, provide sufficient security for nuclear weapons. As noted, PALs were introduced because the two-key system was found to be inadequate to protect U.S. weapons assigned to NATO. Another mechanical device, slightly more complex, is a combination lock—which can be made moderately secure, particularly if no one person knows the entire combination.

An ordinary electrical connecting plug having a large number of pins, at least twenty-four, and perhaps as many as sixty-four, can be used as the key to a very nearly "pick-proof" lock. In this case, particular pairs of pins are wired to a source of specific electrical voltage, while others are left unconnected; keying is accomplished by placing the plug into a connector located in an accessible part of the protected weapons. The connector is attached to a decoding device placed inside the weapon that checks to see that the appropriate pins, and only the appropriate pins, are connected. Such an electronic key offers significant advantages: it depends on relatively primitive technology but, nevertheless, requires a combination as complex as might otherwise be achieved by the use of a computer; the complex electronic key demands far more time to "pick" at random than almost any mechanical lock; and the decoder can be constructed so that it registers any incorrect attempts to unlock the PAL. It might appear that such a keying plug could be rapidly picked by a computer that cycled rapidly through all possible combinations of pins. This can be prevented by incorporating a simple switch that requires the removal and replacement of the plug between tries; at a more sophisticated level, the decoder circuitry could be designed to require at least one second to elapse between tries. If a thirty-two–bit "word" were used as the key, it would then require more than two billion attempts, on average, to find the correct combination by trying all possibilities at random. With one second allowed per try, an average successful attempt would take about sixty-eight years.

According to published descriptions, American PALs now in use come in four different versions, ranked according to their sophistication in categories (CAT) A, B, D, and F.[12] CAT A and B PALs are electromechanical systems incorporating four ten-position selector switches. CAT B systems are designed to be controlled from the cockpit of an airplane and permit authorized users to perform various test and maintenance operations on the PAL by inputting correct codes. CAT D PALs incorporate two additional switches in the coding system as well as the numerical equivalent of master and sub-master keys, allowing a single code to unlock many weapons within a given class (or even across class boundaries). The CAT D PAL was the first device to incorporate devices to limit the number of tries in order to prevent an unauthorized user from searching codes at random. CAT F devices also incorporate the multiple code and limited try capabilities of the CAT D units, but have increased the number of digits in the code to twelve.

---

[12]Thomas B. Cochran, William M. Arkin, and Milton M. Hoenig, *Nuclear Weapons Databook: U.S. Nuclear Forces and Capabilities*, vol. 1 (Cambridge, Massachusetts: Ballinger, 1984), 65.

The earliest American PALs were placed in "inaccessible portions of the weapons." These early PALs apparently worked by interrupting the arming and firing circuitry.[13]

One of the first scientists to advocate the use of permissive action links, Harold Agnew, assessed avoiding bypass of the newly-conceived PALs as the most important consideration for the design of the devices. The PALs were designed so that anyone who attempted to circumvent the mechanism would need "considerable knowledge of information not normally available to non-U.S. forces, or else would require much intuitive guesswork with a high probability of failure."[14]

The governing requirement of the PAL system Agnew proposed in 1961 was that it

> be quite difficult to negate in a short time [hours] without an unusual and unlikely degree of intelligence about the system, even if the U.S. custodians are forcibly removed from effectiveness. While the custodians are performing their job, such negation should be extremely difficult.[15]

These criteria remain important for the design of permissive action links and use-control devices by any nuclear power.

Basic Principles of PAL Design

Nuclear weapons are often stored with one or more essential components, such as their fuses or a portion of the fissile mass, kept separate from the main device. A simple protection mechanism would physically prevent the insertion of the arming component. Conceptually this may be accomplished in two ways: a locked obstacle may be placed on the component making it too large to fit the aperture for its insertion, or an obstruction within the weapon may block the insertion of the component.

The first method is easier to design and build, but offers less protection because the insertable element is smaller than the weapon, more portable, and easier to conceal. Because the locking device is apt to be outside rather than inside the element, an unauthorized person or group may well have an easier time gaining access to the locking mechanism. If the PAL is designed into the element early on, of course, the obstructions— for example, protruding tabs that are either released or retracted when the correct key is inserted—can be placed within the insertable part. Devices that function in this way are not, strictly speaking, PALs in current American usage. They may, however, be quite satisfactory means of preventing unauthorized use of nuclear weapons.

---

[13]Letter from Agnew to Starbird, op. cit., note 7.
[14]Ibid.
[15]Ibid.

Placing the safety mechanism within the nuclear device offers important advantages. The weapon is more massive than the insertable arming/fusing device and is, therefore, more difficult to transport to a place where it can be worked on at leisure. Once again, the simplest protection within the weapon is an obstruction that prevents the insertion of an essential component.

A true PAL, however, must do more than merely prevent arming of a nuclear weapon. It should protect the weapon against all forms of unauthorized use, as well as against tampering, disassembly, and misuse, and must do so even when the weapon is fully assembled, armed, and mounted on its delivery vehicle, ready for quick delivery. The PAL, in fact, must protect the weapon against unauthorized use until the moment that its delivery is ordered by the proper authority—up to the instant at which dispatch of the weapon becomes irrevocable and even to the moment a missile is launched or a bomb is dropped. The PAL also should be able to defeat attempts to examine the weapon and to salvage the finely machined nuclear components.

The PAL decoder can be located somewhere within the arming, fusing, and yield selection control panel. However, a PAL located in an accessible part of the weapon can also be reached by unauthorized personnel who can wire around the PAL if it consists of nothing but a key switch, however sophisticated, in an accessible leg of the firing circuitry. It is much more important, and much more difficult, to prevent bypassing the PAL than to prevent picking the lock. Bypassing a PAL should be, as one weapons designer graphically put it, about as complex as performing a tonsillectomy while entering the patient from the wrong end. For this reason, modern PALs are integral parts buried deeply within the nuclear devices they protect.

Many quite ingenious techniques have been developed to prevent the bypassing of locking circuitry. They cannot be described in publicly available literature, although one widely known technique can serve as an example to demonstrate the general principles used. A switch can be bypassed if an intruder can gain access to the wires leading to and from the lock mechanism. Thus, one challenge is to prevent access to the PAL cabling, which can be done by enclosing the wires themselves in gas-tight pressurized or evacuated tubes. A pressure-sensitive switch can then be placed in the tube, the switch being connected to a small quantity of explosive sufficient to destroy the electrical circuitry needed to detonate the weapon. Any attempt to enter the tube to reach the cabling will cause an immediate change in the air pressure measured by the switch, thus activating it. In turn, this leads to the destruction of the weapon, a drastic act, but presumably preferable to leaving unauthorized personnel in control of a working nuclear weapon.

An unauthorized user might then have to reach the PAL cabling by operating in a pressure chamber so that piercing the tube would not cause the pressure on the switch to drop. Alternatively, the PAL designer can specify the use of evacuated tubing instead, forcing the intruder to don a spacesuit and work in a vacuum. (Alternatively, special pressure-retaining tools could be used, but they are not common. Furthermore, the intruder must know the pressure within the tubing he is attacking in order to set his tools correctly.) Still more effective would be the use of both evacuated and pressurized tubing in the same PAL. These techniques, of course, make servicing the weapons more difficult than would otherwise be the case, a price that must be paid for securing the device against misuse.

Other techniques for reducing the vulnerability of the PAL to tampering might include tilt and motion switches, designed to detect unauthorized removal of the weapon from its normal storage site. These switches, presumably, could be disabled to permit authorized transport of the weapons by the insertion of a special code into the PAL decoder. This code need not be the same as the one that permits the weapon to be used.

An essential ingredient in any package designed to make the PAL tamper resistant is a counter that tallies the number of attempts to insert the code key into the device. Without a limitation on the number of tries, it is at least conceivable that a dedicated person could succeed in obtaining the ability to use a weapon simply by trying codes at random.

For PALs that use an alphabetic code word it is also essential to choose codes that are not plain language words. If ordinary words are used, the temptation to pick something significant or easy to remember (for instance, the name of a spouse or child or an inspirational word such as "Overlord") is often overwhelming, and this can give valuable clues to someone who wishes to defeat the PAL. This problem is familiar to all those who devise computer password schemes and to those who must use computers so protected.

Permissive action links must be designed to meet three fundamental requirements: they must prevent unauthorized use of the weapons by personnel who would, in other circumstances, be the ones to arm and to deliver the atomic ordnance; they must preclude misuse or appropriation of the weapon by intruders—perhaps thieves or terrorists—who would normally be excluded from contact with the weapons; and, finally, PALs must permit the ready use of the weapons when authorized by the national command authorities. Competent engineers and scientists who consider the problem of PAL construction should have no trouble designing and building successful use-control devices that satisfactorily solve the problems of keying and circumvention and so provide enhanced security for their weapons. New nuclear powers should be encouraged to design such devices for their own weapons.

Naval Nuclear Weapons without PALs
Virtually all U.S. land-based nuclear weapons are now equipped with a permissive action link or some other type of coded use-control device.

Nuclear weapons deployed on ships and submarines, however, are not equipped with such devices. In this section, the types of nuclear forces deployed at sea and the use-control policies now in effect are described. Table 1 provides information on naval tactical nuclear weapons and how they are deployed. Table 2 presents the number and type of nuclear strategic missile warheads at sea.

Table 1: Naval Tactical Nuclear Weapons

Weapon Type	U.S.	USSR	U.K.	France	China	Total
Cruise missiles	125	788	0	0	0	913
Aircraft bombs	1,530	0	50	36	130	1,746
Anti-submarine weapons	1,760	1,278	140	0	0	3,178
Anti-air weapons	300	260	0	0	0	560
Naval artillery	0	100	0	0	0	100
Coastal missiles	0	100	0	0	0	100
Total	3,715	2,526	190	36	130	6,597

Table 2: Strategic Missile Warheads at Sea (1987)

U.S.	USSR	U.K.	France	China	Total
5,632	2,902	64	256	39	8,893

Source: *Bulletin of the Atomic Scientists*, September 1987, 63; table adapted from William M. Arkin, "The Nuclear Arms Race at Sea," *Neptune Papers*, no. 1 (Washington, D.C.: Institute for Policy Studies, Greenpeace, 1987).

Together, the United States and the Soviet Union possess more than 6,000 naval tactical nuclear warheads. Great Britain can deploy approximately 50 tactical nuclear bombs at sea and 140 tactical nuclear antisubmarine warfare (ASW) weapons at sea. France can deploy an estimated 36 tactical nuclear bombs and China approximately 130 bombs. In sum, the two superpowers possess 95 percent of the world's tactical nuclear arsenal at sea.[16]

---

[16]Desmond Ball, "Nuclear War at Sea," *International Security*, vol. 10, no. 3 (Winter 1985–86): 23; and William M. Arkin, "The Nuclear Arms Race at Sea," *Neptune Papers*, no. 1 (Washington, D.C.: Institute for Policy Studies and Greenpeace, 1987).

As table 2 shows, there are almost 9,000 strategic missile nuclear war-heads in the world today. About 63 percent of these warheads are deployed on U.S. submarines and 33 percent are based on Soviet submarines. The remaining 4 percent are based on British, French, and Chinese vessels.[17]

The command and control procedures that the Soviet Union exercises over its naval-based tactical nuclear weapons are not altogether clear. However, based on his analysis of more than 260 documents authored by (or issued in the name of) the Soviet navy's commander in chief or the Soviet minister of defense, U.S. Navy Commander James Tritten found "no literature evidence to support the view that release authority for tactical nuclear weapons is a Navy matter nor that a tactical nuclear war at sea alone would be initiated by the Soviets. The decision to initiate tactical nuclear war at sea appears neither a Navy decision nor one that will hinge on Navy matters."[18] Professor Donald C.F. Daniel of the Naval War College has concluded: "There seems to be little prospect that [Soviet] nuclear use at sea will be authorized independently of a decision for use on land, as such authorization would run directly against the grain of Soviet military doctrine since the mid–60s."[19]

Because of the openness of American society, much more is known about the command and control of tactical nuclear weapons in the U.S. Navy. Founded in 1775, the U.S. Navy is one of the oldest institutions in the U.S. government. From 1798 to 1947 the navy was "a wholly autonomous department and service . . . engaged almost exclusively in tasks it set for itself."[20] For most of its history, American naval officers by necessity made decisions without consulting with higher authorities. This reflected the long-standing naval tradition of leaders vesting near-absolute authority in ships' captains at sea. In addition, it reflected the difficulty of communicating between far away ships and Washington. A message from the capital to a ship's captain would often take weeks or even months to reach him. In this environment, independent decisionmaking was an absolute necessity.

With the improvement of communications came greater control over tactical naval operations. During the Cuban missile crisis, for example, President Kennedy and his advisers issued orders directly from the White House to ships participating in the "quarantine" of Cuba. In 1975, President Gerald Ford issued orders directly to U.S. military units that were involved with the rescue of a merchant ship that Cambodians had taken over, the *Mayaguez.*

---

[17]Arkin, op. cit.

[18]Commander James John Tritten, U.S. Navy, *Declaratory Policy for the Strategic Employment of the Soviet Navy,* P–7005 (Santa Monica, California: RAND Corporation, 1984), 210.

[19]Donald C.F. Daniel, "The Soviet Navy and Tactical Nuclear War at Sea," *Survival,* vol. 29, no. 4 (July–August 1987): 333–34.

[20]Richard Neustadt and Ernest May, *Thinking in Time: The Uses of History by Decision-Makers* (New York: Free Press, 1987), 239.

Why the U.S. Navy Opposes PALs on its Vessels

Even with modern communications available, however, many U.S. naval officers contend that there are compelling reasons not to install permissive action links on U.S. ship-based nuclear weapons. First, these weapons are not based on foreign territory and are not, therefore, subject to theft or hostile takeover by foreigners. As noted in the first part of this chapter, one of the strongest motivations for the development of PALs in the early 1960s was the U.S. desire to ensure control over its foreign-based, nuclear-armed forces, particularly quick-reaction-alert aircraft. Tactical nuclear weapons deployed on U.S. Navy ships are based on American "territory" and are controlled by officers in whom the president has placed "special trust and confidence" in the language of the commission every U.S. military officer receives. For that reason, naval officers are prone to regard the use of PALs at sea as indicating a lack of trust in the integrity of naval officers or even to be insulting.

A second reason often given to resist the installation of PALs on sea-based tactical nuclear weapons is that such an installation could seriously compromise the readiness and reliability of the weapons. Retired Vice Admiral Gerald Miller voiced this concern when he wrote: ". . . the real danger may be that so many checks, constraints, and verification procedures are established that the United States will actually be unable to retaliate even with the civilian hierarchy intact, in full control of their faculties and all communications systems in order."[21] Stein and Feaver have described this trade-off between control and vulnerability as the "always/never problem": PALs must always preclude unauthorized detonations and, simultaneously, never interfere with the firing of a weapon if ordered to do so by the proper authorities.[22]

Third, naval officers contend that ship-based nuclear weapons are already safeguarded by a number of stringent procedures. The weapons are equipped with the environmental sensing devices described in the first part of this chapter. An additional security procedure is the "two-man rule" that requires at least two (and in most cases more) officers to authorize and participate in the firing of a nuclear weapon. The U.S. Navy also has an elaborate "personnel reliability program" (PRP) that is designed to preclude anyone with an unstable mental background or a psychological, drug, or alcohol problem from having anything to do with nuclear weapons.

Fourth, U.S. naval authorities in the past have also argued that communications from the national command authority to navy ships are not

---

[21]Vice Admiral Gerald Miller, U.S. Navy ret., "Existing Systems of Command and Control," in Franklyn Griffiths and John Polyani, eds., *The Dangers of Nuclear War* (Toronto: University of Toronto Press, 1979), 59.

[22]Stein and Feaver, *Assuring Control of Nuclear Weapons,* 105–6.

sufficiently reliable to place PALs on tactical nuclear weapons deployed on surface ships.

In sum, the U.S. Navy opposes the installation of PALs on the tactical nuclear weapons deployed on its vessels for four principal reasons: tradition, concern that PALs would reduce reliability and readiness, confidence in existing safety procedures, and lack of confidence in the reliability of communications. Given their tradition and operational responsibilities, it is understandable that naval officers prefer to have independent control over the nuclear weapons on their ships. The conflicting interest, however, is that of the broader society that naval officers serve: to ensure maximum control over nuclear weapons to minimize the possibility of accidental or unauthorized use.

Reasons to Place PALs on Naval Nuclear Weapons

Because of the tremendous damage that can be inflicted by nuclear weapons, permissive action links have been installed on nuclear weapons manned by army and air force officers. The tradition of naval autonomy should bow to change as well, for there are compelling reasons to do so. Furthermore, naval nuclear weapons cannot at all times be assumed to be in the absolute control of U.S. forces. The use of PALs on naval weapons, if technically feasible, should be viewed as a prudent measure and should not be interpreted by naval officers as in indication of mistrust.

Several events and some personnel statistics raise serious questions about the effectiveness of the military's personnel reliability program. There have been cases in which military leaders acted on their own with no authorization from higher authorities for their actions. During the Vietnam War, for example, Air Force General John Lavelle ordered twenty unauthorized bombings of North Vietnam in 1971–72. More recently, a former navy warrant officer, John Walker (who was at one time stationed on a U.S. nuclear submarine), his brother (a commissioned naval officer), his son (a navy enlisted man), and a close friend (a chief petty officer) sold some of the navy's most sensitive secrets, including the operational orders of submarines and cryptographic information, to the Soviet Union over a period of many years.[23]

Approximately 114,000 to 118,000 people within the Department of Defense have nuclear weapons responsibilities.[24] Most of them are military personnel performing their duties at approximately 50 U.S. air bases supporting strategic bombers, missiles, and naval combatants, and at about 160 bases in Europe and other parts of the world supporting theater

[23]John Barron, *Breaking the Ring* (Boston: Houghton Mifflin Company, 1987); and Howard Blum, *I Pledge Allegiance . . . The True Story of the the Walkers: An American Spy Family* (New York: Simon and Schuster, 1987).
[24]Cochran, Arkin, and Hoenig, *Nuclear Weapons Databook,* 84.

nuclear forces.[25] All of those who work with nuclear weapons must be cer-
tified by the military's personnel reliability program. About half of the
personnel in this program are from the air force. Of this number, 3,093
(5.6 percent) have been decertified and removed from their jobs for reasons that included drug or alcohol abuse and/or mental, disciplinary, or
physical problems.[26] In 1981 6.9 percent were removed, and in 1982
6.3 percent were removed from the program. Former Assistant Secretary
of Defense Donald R. Cotter has pointed out that the PRP ". . . is designed
to prevent the assignment of unreliable persons to nuclear duties through
a screening process and then to remove from nuclear duties those persons whose reliability, trustworthiness, and dependability become incon-
sistent with standards. . . . No initial screening process can guarantee
future behavior. . . ."[27] The Walker spy ring is disturbing proof that the
PRP is not perfect.

During the past two decades, the U.S. Navy has grown increasingly
dependent on advanced communications. At the present time, naval communications have advanced to the point where naval analyst Karl
Läutenschläger contends: "We are probably now in the midst of the next
major transition [in the evolution of naval technology]."[28] By the end of
this decade, the U.S. Navy will be dependent on satellites for data collection, targeting information, navigation, and early warning.[29]

Those who are opposed to placing PALs on naval nuclear weapons
argue that if PALs were placed on sea-based nuclear weapons, their use
could be precluded by the effective employment of electronic warfare by
the enemy. There are four reasons to discount this argument. First, military planners are justifiably concerned about electronic countermeasures
and, because of this, a great deal of research has gone into ensuring that
communications will be maintained with military units during peace and
war. Second, the navy has developed an anti-jam mode of communications. In this mode, a message is transmitted in such a way that it can be
deciphered even if only parts of it are received. Third, as Walter Slocombe
has pointed out, "the critical requirement of any transmission system is
that the outward links of communications be sufficiently redundant to
prevent effective disruption. Because only one channel to each weapon
need work—the great strength of the transmission system—virtually every
vulnerability would have to be exploited simultaneously to disrupt the

[25]Cotter in Carter, Steinbruner, and Zraket, *Managing Nuclear Operations,* 18.
[26]Center for Defense Information, "Who Could Start a Nuclear War?" *The Defense Monitor,*
vol. 14, no. 3 (1985): 4. The figures cited are for the air force.
[27]Donald R. Cotter quoted in ibid.
[28]Karl Läutenschläger, "Technology and the Evolution of Naval Warfare," *International Security,* vol. 8, no. 2 (Fall 1983): 44.
[29]Louise Hodgden, "Satellites at Sea: Space and Naval Warfare," *Naval War College Review,*
vol. 32, no. 4 (July–August 1984): 43.

whole system."[30] The fourth reason that one need not be unduly concerned about electronic warfare concerns the uncertainty of electronic counter-measures: ". . . electronic vulnerabilities are usually partial rather than total, sometimes temporary, and always uncertain both for the attacker and for the user of the C³I system's electronic equipment."[31]

At the present time U.S. Navy ships must receive an "emergency action message" from the national command authority in order to fire nuclear weapons. If officials are confident that this message can be received, then why could not a PAL code also be securely transmitted and received? As Ashton Carter has pointed out, "If the [PAL] codes are short and simple, as they could be, they can easily accompany orders authorizing use of the weapons."[32] An effective PAL code should require no more than two computer "words" (thirty-two or sixty-four bits each).

It is certainly possible than an enemy would try to disrupt U.S. communications either through jamming or direct attacks on the satellites that are used to relay much of U.S. communications traffic. As discussed earlier, total disruption of *outward* communications is both difficult and chancy for an opponent. As a result, the United States is confident that its communications systems are currently adequate in both peace and war. If this is so, then the uncertainty of communications to surface ships is not, itself, justification for rejecting the use of PALs at sea. Nuclear weapons may not be used without direct authorization, and so it is plausible to include a PAL code in the action message directing their employment.

PALs for Submarine-Launched Missiles?

Not a great deal is publicly known about the procedures that various countries employ to command and control these systems. From available evidence, it appears that French nuclear forces must receive two authenticated codes before an SLBM (submarine-launched ballistic missile) can be enabled. One American nuclear weapons expert who is familiar with U.S. use-control procedures and technology visited a French ballistic missile submarine and was reportedly impressed with the security of the SLBMs in it.[33] The command and control of British SLBMs are based on the "two-man rule" and are similar to the U.S. procedures for its SLBMs.[34]

---

[30]Walter Slocombe, "Preplanned Operations," in Carter, Steinbruner, and Zraket, *Managing Nuclear Operations*, 138.

[31]Ashton B. Carter, "Communications Technologies and Vulnerabilities," in Carter, Steinbruner, and Zraket, op. cit., 280–81.

[32]Ashton B. Carter, "Assessing Command System Vulnerability," in Carter, Steinbruner, and Zraket, op. cit., 593.

[33]Stein and Feaver, *Assuring Control of Nuclear Weapons,* note 3, 87.

[34]Shaun Gregory, *The Command and Control of British Nuclear Weapons,* Peace Research Report No. 13 (Bradford, United Kingdom: School of Peace Studies, University of Bradford, December 1986), 72–77.

According to press accounts, to fire the missiles from U.S. ballistic missile submarines,

> . . . a message from the President must come over one of several channels of communication and be authenticated by at least two officers. Then it must be further authenticated by the executive officer and the captain. After the captain orders the crew to begin preparations for firing, he must obtain launching keys from a safe to which he does not have the combination. Launching the missile takes four keys turned by four different officers, two in the missile control room, another at the gas generator two decks below that propels the four-story high missile out of the tube, and the last by the captain in the command center. Two more officers must open that, one the outside door and the other the inside door.[35]

The principal reason that PALs are not installed on U.S. submarine weapons is the difficulty in communicating with submerged vessels. In fact, because of the difficulties in communicating from surface ships to submarines, it was not until 1970 that the U.S. Navy began having submarines operate with carrier task forces, and ship-to-submarine tactical communications are much less demanding than long-range communications with submarines. If a submarine surfaces or even remains submerged and raises an antenna above the surface in order to receive communications, the possibility of detection increases. Thus, the United States has turned to four principal modes for communicating with submarines: broadcasting a high frequency (HF) radio message from ships and land stations, "Take Charge and Move Out" (TACAMO) messages broadcast by aircraft, and ground-based transmitters employing very low frequency (VLF) and extremely low frequency (ELF) radio. VLF can broadcast at about the same rate as a teletype machine. ELF is much slower; it can transmit about one bit per minute, which means that a standard emergency action message for land-based forces would require up to thirty minutes to receive, even without including a PAL code.[36] ELF has some advantages over VLF because the submarine can remain deeply submerged while receiving an ELF-generated message. To receive a VLF message, the submarine cannot submerge deeply and must trail an antenna within the upper ten

---

[35]Richard Halloran, "Submarine's Patrol Sets Silent, Random Course," *New York Times,* December 6, 1987, A–25; see also Ball, "Nuclear War at Sea," 10–11; Jim Bencivenga, "Aboard a Nuclear Sub—A World of Computers, Sonar, Silence, and Stealth," *Christian Science Monitor,* October 14, 1982; and Lawrence Meyer, "AF Locks System Urged for Navy's Nuclear Missiles," *Los Angeles Times,* October 14, 1984, 28.

[36]Carter, "Communications Technologies and Vulnerabilities," in Carter, Steinbruner, and Zraket, *Managing Nuclear Operations,* note 5, 235–36.

meters or so of the ocean where the VLF waves penetrate. Alternatively, the submarine can trail a long length of wire so that a hundred meters of the wire lie near the surface. When this is done, the diving depth and speed of the submarine are limited. In addition, there are reports that the wire sometimes floats on the surface, providing an indication of the location of the submarine. According to the FY 1989 Department of Defense Posture Statement, the ELF transmitter site in Michigan will become operational in 1989, and the entire ELF submarine communication system will become fully operational in 1990 when receivers have been placed on all ballistic missile submarines.[37]

Communications have advanced to the point that permissive action links can be placed on the nuclear weapons deployed on naval surface warships and submarines. The U.S. Navy is already heavily dependent on communications and has invested billions of dollars to ensure that its communications will be effective even in wartime. According to present policy, naval commanders must receive an emergency action message in order to use the nuclear weapons under their command. Adding a permissive action link code to this message does not constitute an overly stringent requirement. Not requiring a PAL code for naval nuclear weapons only makes sense if the U.S. government is no longer going to require naval commanders to receive authorization prior to releasing nuclear weapons—a change in U.S. policy that few people advocate.

Several steps could be taken to ensure that the emergency action message and PAL code will be received by naval ships and submarines. First, the two TACAMO communications planes could be supplemented with several additional planes.[38] Second, deployed ballistic missile firing submarines (SSBNs) could monitor communications at more frequent intervals. Third, in the event of a crisis, transmitters could be deployed to transmit messages to submarines through the deep sound channel. Fourth, fiber optic systems with acoustic nodes could be deployed on the ocean floor. Submarines could "plug into" these nodes in order to receive messages. If these measures were taken, communications would be highly redundant and any enemy of the United States could not be certain of blocking U.S. communications with its forces.

Policy Proposals for Permissive Action Links

There are at least four ways the U.S. government could help to reduce the risk of the misuse of nuclear weapons by promoting the use of PALs.

---

[37]Frank C. Carlucci, *Annual Report to the Congress, Fiscal Year 1989* (Washington, D.C.: GPO, 1988), 241.

[38]Bruce G. Blair, "Alerting in Crisis and Conventional War," in Carter, Steinbruner, and Zraket, *Managing Nuclear Operations*, 98.

1. The U.S. government could provide information about PALs to other states.[39]
   There have been reports that in the early 1960s the United States
provided information about permissive action links to Great Britain and
the USSR.[40] According to these reports, American leaders at the time assumed that centralized control over nuclear weapons by the governments
that owned them would reduce the risk of accidental and unauthorized
launch of nuclear weapons. Additionally, such a sharing of PAL information could diminish concerns that nuclear weapons might be used in times
of domestic turmoil or international crisis, thus diminishing pressures
for preemptive attacks, particularly in conflict-prone areas of the world
such as South Asia. India has tested a nuclear device and Pakistan is believed to be close to developing nuclear weapons. If Indian and Pakistani
leaders each knew that the nuclear weapons of both countries were
equipped with PALs, then the danger of inadvertent escalation and pre-
emption in crises would be reduced. In addition, the danger of dissident
or separatist groups gaining control over nuclear weapons would be
reduced if PALs were installed on any nuclear weapons that might exist
in South Asia or in other areas where the overthrow of a central government is a distinct possibility.
   Still, there may be significant potential disadvantages to providing
other countries with information about permissive action links. First, there
is a real danger that such a transfer of information would encourage states
to develop nuclear weapons; if control were assured, why not go ahead
and develop nuclear weapons? At a minimum, it might appear that the
United States was encouraging proliferation, thus undercutting its efforts
to slow the spread of nuclear weapon capabilities. Yet most experts do
not believe that nuclear proliferation can be stopped. The question most
often discussed today is not *whether* but *when* additional states will "go nuclear."[41] A state's decision to develop nuclear weapons, however, will not
hinge on whether or not PALs are available; rather, the decision will be
based on leaders' perception of the political and military value of pos-
sessing nuclear weapons. If proliferation is going to occur in any event,
then it is better to make it happen under relatively safe, rather than relatively more dangerous, conditions.
   A second concern is that providing information about PALs to other
countries would compromise U.S. nuclear weapons security. If the United

[39]This is one of the recommendations of the International Task Force on the Prevention
of Nuclear Terrorism; see Paul Leventhal and Yonah Alexander, eds., *Preventing Nuclear Terrorism: The Report and Papers of the International Task Force on Prevention of Nuclear Terrorism* (Lexington, Massachusetts: Lexington Books, 1987), 39.
[40]Edward Klein and Robert Littell, "SHH! Let's Tell the Russians," *Newsweek,* May 5, 1969,
46–47.
[41]Lewis Dunn, *Controlling the Bomb: Nuclear Proliferation in the 1980s* (New Haven, Connecti-
cut: Yale University Press, 1983).

States were to give other countries information about currently deployed PALs, those countries (or others) could gain a better idea of how these links work and possibly some idea of how to defeat them. One way around this concern is to provide only the most basic information about PALs. Indeed, the principles of PAL operation have already been discussed in this chapter without providing useful detail on the engineering principles of operational PALs.

Information could be made available to other countries in a number of different ways. The U.S. government could publish an unclassified paper that describes the basic design and technical details of permissive action links. Such a paper could be made available from the National Technical Information Service to anyone who requested it at a nominal cost. The U.S. government or a foundation could sponsor a conference focusing on the technical aspects of PALs. The conference on "Assuring Control of Nuclear Weapons" sponsored by Harvard's Center for Science and International Affairs in February 1986 focused on the history of the development of PALs. A follow-on conference might focus on the technical aspects.[42]

2. The U.S. government could take the initiative to provide rudimentary permissive action link devices to other countries.

This proposal is merely a strengthened variant of the proposal above.[43] The major advantage is that the recipient country would obtain a tested device that worked. Given the suspicion and hostility toward the United States that exists in many countries, it is possible that U.S. permissive action links would not be accepted even if offered. But those countries that accepted the U.S. offer through "reverse engineering" could develop their own PAL devices. It is possible, however, that others could gain a better understanding of U.S. permissive action links and, therefore, U.S. nuclear weapons security. Again, the way around this danger is for the United States to provide basic devices and not currently deployed models.

3. The U.S. government should install permissive action links on nuclear weapons deployed on all U.S. Navy surface ships.

Because these weapons are not presently secured by PALs, they can, in fact, be armed and fired without authorization. Communications with surface ships are today both reliable and secure; indeed, surface ships must receive an "emergency action message" authorizing the use of nuclear weapons. Because communications with surface ships are also fast, inclusion of a permissive action link code with the emergency action message

---

[42]For a report on this conference, see Stein and Feaver, *Assuring Control of Nuclear Weapons.*
[43]This is also a recommendation of the International Task Force on the Prevention of Nuclear Terrorism; see Leventhal and Alexander, *Preventing Nuclear Terrorism*, note 39.

poses no serious operational problem for the commander of a nuclear-armed surface vessel. There are three reasons to adopt such a procedure.

First, in recent years, the U.S. Navy has developed and adopted a new "Maritime Strategy" that calls for the navy to use "early, forceful, global, forward deployment of maritime power both to deter war with the Soviet Union and to achieve U.S. war aims should deterrence fail."[44] In operational terms, the new strategy calls for navy attack submarines (SSNs) at the beginning of a crisis to move into the Barents Sea and, if called upon, to seek out and destroy both Soviet attack submarines and SSBNs and surface ships and submarines equipped with cruise missiles.[45] The Maritime Strategy also calls for U.S. carrier–based aircraft to attack targets in the Soviet Union, if ordered to do so. Both missions would require large numbers of U.S. warships to operate in waters proximate to Soviet territory.

Under these conditions, it is vital for American decisionmakers to be able to control U.S. Navy forces to the maximum extent possible. If the highest political authorities in the United States do not have assured control over U.S. naval forces, there is a possibility, however remote, that American naval commanders could independently use the nuclear weapons deployed on board their ships and submarines. And the Maritime Strategy makes deterrence less stable. The installation of permissive action links on nuclear weapons aboard U.S. Navy ships would increase central control by U.S. civilian authorities over naval nuclear weapons and decrease the risk that a crisis would lead to war.

Second, the U.S. Navy currently plans to buy 3,994 Tomahawk sea-launched cruise missiles through the 1990s to deploy on 198 ships and submarines. Of these missiles, 758 will have nuclear warheads, an increase of perhaps 25 percent in the number of naval nuclear weapons.[46] It can, of course, be argued that the deployment of nuclear-armed Tomahawks in no way changes the picture because most Tomahawks will be deployed on ships that already carry tactical nuclear weapons. But we disagree with this view because the land-attack Tomahawk has the potential to carry nuclear war from the sea to the shore, a qualitative change of great significance. This increased risk is unacceptable given improvements in communications technology.

---

[44]Captain Linton F. Brooks, U.S. Navy, "Naval Power and National Security: The Case for the Maritime Strategy," *International Security*, vol. 11, no. 2 (Fall 1986): 58; see also Admiral James Watkins, "The Maritime Strategy," *U.S. Naval Institute Proceedings 112* (January 1986), 3–17.

[45]Barry R. Posen, "Inadvertent Nuclear War? Escalation and NATO's Northern Flank," *International Security*, vol. 7, no. 2 (Fall 1982): 28–54; John J. Mearsheimer, "A Strategic Misstep: The Maritime Strategy and Deterrence in Europe," *International Security*, vol. 11, no. 2 (Fall 1986): 3–57; George Quester, "Maritime Issues in Avoiding Nuclear War," *Armed Forces and Society*, vol. 13, no. 2 (Winter 1987): 189–204; and George Quester, "Through the Nuclear Strategic Looking Glass, or Reflections Off the Window of Vulnerability," *Journal of Conflict Resolution*, vol. 31, no. 4 (December 1987): 725–37.

[46]*Bulletin of the Atomic Scientists*, December 1987, 51.

Communications with naval surface vessels are currently reliable, redundant, and robust enough to provide confidence that PAL codes could be received along with the emergency action message authorizing the use of nuclear weapons. If communications with surface vessels are not reliable enough to provide confidence in the receipt of an emergency action message including a PAL code, then the system is not adequate to support naval warfare, period. If that is the case, then the full attention of the naval procurement system must be directed to obtaining reliable and secure communications between naval vessels and the Pentagon. Unless it is intended that commanders at sea should have independent authority to cross the nuclear threshold, placing PALs on naval nuclear weapons in no significant way limits the freedom of action of those captains.

Third, there are indications that the U.S. Navy, perhaps as a part of the new Maritime Strategy, is moving in the direction of greater autonomy and independence. Illustrative of this trend is the following statement from an essay that won the Naval War College's Admiral Richard G. Colbert Memorial Prize.[47] According to U.S. Navy Lieutenant Commander T. Wood Parker,

> Our current release procedures and rules are reactive in nature and restrict considerably the actions of commanders at sea. These restrictions could be critical in hostilities, particularly if the Soviet Navy decided to use tactical nuclear weapons against our ships. Without going into detail it appears that increased decentralization, flexibility, and even leniency may be required. One hopes study of this matter will begin soon.[48]

Recent developments—the Maritime Strategy and the imminent deployment of hundreds of nuclear Tomahawk cruise missiles—call for more, rather than less, control over these forces. To the extent that this chapter may reflect the thinking of senior naval officials, it strengthens the case for the inclusion of PALs on naval tactical nuclear weapons.

4. The U.S. government should install permissive action links on all submarine-launched nuclear weapons.

The arguments in favor of installing PALs on all submarine-based nuclear weapons are essentially the same as those presented for surface ships. In evaluating this proposal, however, there is a crucial variable to consider: communications. Using present technologies, it is more difficult

---

[47]Editor, *Naval War College Review*, vol. 35, no. 1 (January–February 1982): 3.

[48]Lieutenant Commander T. Wood Parker, U.S. Navy, "Theater Nuclear Warfare and the U.S. Navy," Admiral Richard G. Colbert Memorial Prize Essay, *Naval War College Review*, vol. 35, no. 1 (January–February 1982): 15.

to communicate with submerged submarines in a timely, reliable manner than with surface ships. Communication with submarines is currently maintained via several different means, as previously discussed in this chapter.

Some authorities argue that ELF technologies are too slow to transmit and receive an emergency action message and a PAL release code. The launch of SLBMs in retaliation for a first strike against American territory need not be carried out on a timetable delineated in seconds unless the naval warheads are directed at unused Soviet ICBMs, so even relatively slow communications should suffice to transmit the action message. Consider, however, the case in which SLBMs are to be used to attack missiles in their silos.

If Trident D–5 warheads are to be used for such a damage-limiting role, the action message to each submarine must designate the precise Soviet silos that are to be attacked and, probably, the specific warheads and missiles that are to be used against each one. Assuming that the locations of all Soviet missile silos are already stored on board the submarine, the action message must at least list an identification number for each target. Because the Soviets currently have 1,400 ICBMs, each identification number must consist of at least eleven bits (binary coding, $2^{11} - 1 = 2047$). If a submarine is ordered to expend twelve missiles attacking 120 targets, the message must be at least 120×11 bits long, exclusive of identifying and authenticating codes, launch times, address blocks, and so on. The addition of a thirty-two–bit PAL code to such a message clearly is not a serious burden and probably requires not more than an additional minute of transmission and reception time. It is, however, a message that must be transmitted using VHF or VLF channels—not ELF—particularly if the weapons from several ships are to be coordinated. If VHF or EHF communications are available, the delay caused by the inclusion of a PAL code in the release order is a small fraction of a minute. In none of these cases is submarine vulnerability significantly increased from its current level.

If ELF channels are the only surviving means for contacting submarines at sea, then it is impossible to execute a prompt retaliation in any case. Even at a communications rate of one bit per minute, the addition of a ten–bit suffix to the emergency action message would not seriously compromise the mission of the submarine receiving the orders. Conceptually, one can even consider an emergency action message consisting of thirty-two bits, transmitted in a cipher changed at least daily, which decrypts as the order to fire with instructions that the entire message (perhaps multiplied by a key changing frequently) be inserted into each missile as the PAL code. In this situation the emergency action message is, in effect, the PAL code, and neither the submarine nor the transmitting authority is in any way burdened by the use of PALs on SLBMs. If such a message cannot be transmitted and received reliably in all circumstances, even in the event of jamming, then the communications system itself must be restructured.

What options are foreclosed by requiring PALs on SLBMs? Only one: the opportunity for a submarine commander, wholly cut off from communications with the National Command Authority, to take nuclear retaliation into his own hands. Despite a great respect for the commanders of U.S. missile submarines, this is an option that ought to be foreclosed in peacetime and at all levels of conflict below the nuclear threshold. It could, of course, be resurrected in extremis by pre-delegating nuclear authority to submarine commanders at sea and transmitting the PAL codes shortly before the outbreak of strategic warfare.

In sum, given current communications technologies and capabilities, communication with submarines can be characterized as reliable and redundant. By placing PALs on submarine-based nuclear weapons, stability is greatly increased at the cost of reducing the independent action of naval commanders.

Several steps could be taken in the future to increase confidence that communications with submarines can be maintained. The current two TACAMO aircraft—one in the Pacific and one in the Atlantic—could be augmented with two additional aircraft.[49] Blue-green visible light is the only band of the electromagnetic spectrum other than VLF and ELF that penetrates seawater, and research is being conducted on the possibility of using eximer blue lasers positioned on submarine laser communications satellites to improve communications with submarines.[50] Research is also being conducted on the possibility of using acoustic signals to transmit messages to submarines.[51] A rudimentary type of communication would be to explode depth charges as a "bell ringer" alerting submarines that something important is happening and that they should monitor their available communications channels closely, including perhaps rising closer to the surface so that VLF or HF antennas could be extended to receive a message. The United States could also build a number of communication nodes linked by fiber optic cables on the ocean floor. The U.S. Navy has extensive experience with a similar antisubmarine warfare system, the Sound Surveillance System, which is a passive system for detecting and tracking submarines. A new system would have the capability to transmit and not simply to listen.

---

[49]Blair in Carter, Steinbruner, and Zraket, *Managing Nuclear Operations*, note 38, 98.

[50]Albert Wohlstetter and Richard Brody, "Continuing Control as a Requirement for Deterring," in Carter, Steinbruner, and Zraket, *Managing Nuclear Operations*, 169.

[51]Tom Stefanick, "The Nonacoustic Detection of Submarines," *Scientific American*, vol. 258, no. 3 (March 1988): 41–47.

Conclusion

Government policies can be evaluated on the basis of two basic criteria: desirability and feasibility. Measures that reduce the risk of accidental or unauthorized detonation of nuclear weapons without reducing a state's own national security are desirable. This is the case with regard to the sharing of PAL technology with other nations, particularly newly emerging nuclear powers. A variety of means are available for the U.S. government to share some PAL technology without compromising the security or design of its own nuclear forces. These means include the open publication of a technical paper describing PALs that provide information on PALs to selected countries and/or sponsorship of a conference focusing on the technical aspects of permissive action links.

When it became feasible to place permissive action links on land- and aircraft-based nuclear weapons, the United States did so. Communications with surface naval vessels have advanced to the point where it is now feasible to install PALs on weapons deployed on these vessels. Although communications with submarines are slower than with surface ships, it is also both desirable and feasible to place PALs on submarine-based nuclear weapons.

APPENDIX
ACCIDENTS INVOLVING U.S.
NUCLEAR WEAPONS, 1950–80[1]

The following unclassified summaries describe the circumstances surrounding thirty-two accidents involving nuclear weapons. Despite the very severe stresses imposed upon the weapons involved in these accidents, there never has been even a partial inadvertent U.S. nuclear detonation. All detonations reported in these summaries involved conventional high explosives (HE) only. Only two accidents, those at Palomares and Thule, resulted in a widespread dispersal of nuclear materials.

Nuclear weapons have never been carried on training flights. Most of the aircraft accidents represented here occurred during logistic and ferry missions or airborne alert flights by Strategic Air Command (SAC) aircraft. Airborne alert was terminated in 1968 because of:

> accidents, particularly those at Palomares and Thule; the rising cost of maintaining a portion of the SAC bomber force constantly on airborne alert; and, the advent of a responsive and survivable intercontinental ballistic missile force that relieved the manned bomber force of a part of its more time-sensitive responsibilities. (A portion of the SAC force remains on nuclear *ground* alert.)

Since the location of a nuclear weapon is classified defense information, it is Department of Defense policy normally neither to confirm nor deny the presence of nuclear weapons at any specific place. In the case of an accident involving nuclear weapons, their presence may not have been divulged at the time depending upon the possibility of public hazard or alarm. For diplomatic reasons, the location of the accidents that occurred overseas, except for Palomares and Thule, has not been specified.

Most of the weapon systems involved in these accidents—including the B–29, B–36, B–47, B–50, B–58, C–124, F–100, and P–5M aircraft, and the Minuteman I missile—are no longer in active inventory. With some

---

[1]Source: Unclassified document released by Departments of Defense and Energy, April 1981.

early models of nuclear weapons, it was standard procedure during most operations to separate the capsule of nuclear material from the weapon for safety purposes. While a weapon with the capsule removed might still contain a quantity of natural (not enriched) uranium, the extremely low radioactivity of this material assumed that an accidental detonation of the HE element would not cause a nuclear detonation or dispersal of toxic materials. More modern weapon designs incorporate improved redundant safety features to insure that a nuclear explosion does not occur as the result of an accident.

The events outlined below involved operational weapons, nuclear materials, aircraft, and missiles under control of the U.S. Air Force, U.S. Navy, or a Department of Energy predecessor agency, the Atomic Energy Commission. The U.S. Army has never experienced an event serious enough to warrant inclusion in a list of accidents involving nuclear weapons. The U.S. Marine Corps does not have custody of nuclear weapons in peacetime and has experienced no accidents or significant incidents involving them.

Because reporting requirements varied among the services, particularly in the earlier period covered by these narratives, it is possible that an earlier accident has gone unreported here. All the more recent events (up to April 1980), however, have been evaluated and included if they fall within the following operational definition of an accident.

The Departments of Defense and Energy have defined an "accident involving nuclear weapons" broadly as an unexpected event involving nuclear weapons or nuclear weapons components that results in any of the following:

- accidental or unauthorized launching, firing, or use, by U.S. forces or supported allied forces, of a nuclear-capable weapon system that could create the risk of an outbreak of war;
- nuclear detonation;
- non-nuclear detonation or burning of a nuclear weapon or radioactive weapon component, including a fully assembled nuclear weapon, an unassembled nuclear weapon, or a radioactive nuclear weapon component;
- radioactive contamination;
- seizure, theft, or loss of a nuclear weapon or radioactive nuclear weapon component, including jettisoning; public hazard, actual or implied.

Accidents Involving U.S. Nuclear Weapons or their Components
1. FEBRUARY 13, 1950. PACIFIC OCEAN, OFF COAST OF BRITISH COLUMBIA.

After six hours of flight, a B-36 enroute from Eilson Air Force Base (AFB) to Carswell AFB on a simulated combat profile mission developed serious mechanical difficulties, making it necessary to shut down three engines. The aircraft was at an altitude of 12,000 feet; the weapon aboard the aircraft had a dummy capsule installed. Icing conditions complicated the emergency and level flight could not be maintained. The aircraft headed out over the Pacific Ocean and dropped the weapon from 8,000 feet. A bright flash occurred on impact, followed by a sound and shock wave. Only the weapon's high explosive (HE) material detonated. The aircraft was then flown over Princess Royal Island where the crew bailed out. The aircraft wreckage was later found on Vancouver Island.

2. APRIL 11, 1950. MANZANO BASE, NEW MEXICO.

A B-29 containing both a weapon and its nuclear capsule departed Kirtland AFB at 9:38 p.m. and crashed into a mountain on Manzano Base approximately three minutes later killing the crew. Detonators were installed in the bomb on board the aircraft. The bomb case was demolished and some HE material burned in the gasoline fire. Other pieces of unburned HE were scattered throughout the wreckage. But because the capsule had not been inserted in the weapon, a nuclear detonation could not have occurred. Four spare detonators in their carrying case were recovered undamaged; the recovered components of the weapon were returned to the Atomic Energy Commission. There were no contamination problems.

3. JULY 13, 1950. LEBANON, OHIO.

A B-50 on a training mission from Biggs AFB, Texas, nosed down and flew into the ground killing four officers and twelve airmen. The high explosive portion of the weapon aboard detonated on impact. There was no nuclear capsule aboard the aircraft.

4. AUGUST 5, 1950. FAIRFIELD SUISUN AFB, CALIFORNIA.

A B-29 carrying a weapon, but no capsule, lost two propellers and experienced landing gear retraction difficulties on takeoff from Fairfield-Suisun AFB (now Travis AFB). The aircraft crashed and burned when the crew attempted an emergency landing. The fire was fought for twelve to fifteen minutes before the weapon's HE material detonated. Nineteen crew members and rescue personnel were killed in the crash and the resulting detonation, including General Travis.

5. NOVEMBER 10, 1950. OVER WATER, OUTSIDE UNITED STATES.

Because of an in-flight aircraft emergency, a weapon not containing a capsule of nuclear material was jettisoned from a B-50 over water from an altitude of 10,500 feet. A high-explosive detonation was observed.

6. MARCH 10, 1956. MEDITERRANEAN SEA.

A B-47 enroute from MacDill AFB to an overseas air base disappeared in heavy cloud cover as it prepared for mid-flight refueling. An extensive search failed to locate any traces of the missing aircraft or crew. No weapons were aboard the aircraft, only two capsules of nuclear weapons material in carrying cases. A nuclear detonation was not possible.

7. JULY 27, 1956. OVERSEAS BASE.

A B–47 aircraft with no weapons aboard was on a routine training mission making a touch-and-go landing when the aircraft suddenly slid off the runway, crashing into a storage igloo containing several nuclear weapons. The bombs did not burn or detonate. There were no contamination or cleanup problems. The damaged weapons and components were returned to the Atomic Energy Commission. The weapons that were involved were in a storage configuration; no capsules of nuclear materials were in the weapons or present in the building.

8. MAY 22, 1957. KIRTLAND AFB, NEW MEXICO.

A B–26 was ferrying a weapon from Biggs AFB, Texas, to Kirtland AFB when the weapon dropped from the bomb bay taking the bomb bay doors with it. Weapon parachutes were deployed but apparently did not fully retard the fall because of the low altitude. The impact point was approximately 4.5 miles south of the Kirtland control tower and .3 miles west of the Sandia Base reservation. The HE material detonated, completely destroying the weapon and making a crater approximately twenty-five feet in diameter and twelve feet deep. Fragments and debris were scattered as far as one mile from the impact point. The release mechanism's locking pin was being removed at the time of release. (It was standard procedure to remove the locking pin during takeoff and landing to allow for emergency jettison of the weapon.)

Recovery and cleanup operations were conducted by Field Command, Armed Forces Special Weapons Project. Radiological survey of the area disclosed a radioactivity level of 0.5 milliroentgen at the lip of the crater but no radioactivity beyond the crater's edge. There were no health or safety problems. Although the weapon and capsule were on board the aircraft, the capsule was not inserted and nuclear detonation could not have taken place.

9. JULY 28, 1957. ATLANTIC OCEAN.

Two weapons were jettisoned from a C–124 aircraft off the east coast of the United States. There were three weapons and one nuclear capsule aboard the aircraft at the time. Nuclear components were not installed in the weapons. The C–124 aircraft was enroute from Dover AFB, Delaware, when a loss of power occurred in the first and second engines. While maximum power was applied to the remaining engines, level flight could not be maintained. At this point, the decision was made to jettison cargo to ensure aircraft and crew safety. The first weapon was jettisoned at an altitude of 4,500 feet; the second at approximately 2,500 feet. No detonation occurred from either weapon. Both weapons are presumed to have been damaged by impact with the ocean surface. The C–124 landed at an airfield in the vicinity of Atlantic City, New Jersey, with the remaining weapon and the nuclear capsule aboard. A search for the weapons or debris proved fruitless.

10. OCTOBER 11, 1957. HOMESTEAD AFB, FLORIDA.

A B–47 departed Homestead AFB shortly after midnight on a deployment mission. Shortly after liftoff one of the aircraft's outrigger tires exploded. The aircraft crashed in an uninhabited area approximately 3,800 feet from the end of the runway. The aircraft was carrying one weapon in ferry configuration in the bomb bay and one nuclear capsule in a carrying case in the crew compartment. The weapon was enveloped in flames, burned and smoldered for approximately four hours until it was cooled with water. Two small HE detonations occurred during the burning. The nuclear capsule and its carrying case were recovered intact, slightly damaged by heat. Approximately one-half of the weapon remained. All major components were damaged but were identifiable and accounted for.

11. JANUARY 31, 1958. OVERSEAS BASE.

A B–47 with one weapon in strike configuration was making a simulated takeoff during an exercise alert. When the aircraft reached approximately thirty knots on the runway, the left rear wheel casting failed. The tail struck the runway and a fuel tank ruptured. The aircraft caught fire and burned for seven hours. Firemen fought the fire for ten minutes, then evacuated the area, fearing an explosion. The high explosive did not detonate, but there was some contamination in the immediate area of the crash. After the wreckage some asphalt was removed and the runway washed down; no contamination was detected. Following the accident, exercise alerts were temporarily suspended and B–47 wheels were checked for defects.

12. FEBRUARY 5, 1958. SAVANNAH RIVER, GEORGIA.

A B–47 on a simulated combat mission collided with an F–86 aircraft at 3:30 a.m. near Savannah, Georgia. The B–47 attempted three times to land at Hunter AFB, Georgia, with a weapon aboard. Because the aircraft could not be slowed enough to insure a safe landing, the decision was made to jettison the weapon rather than expose Hunter AFB to the possibility of a high-explosive detonation. A nuclear detonation was not possible since the nuclear capsule was not aboard the aircraft. The weapon was jettisoned into the water several miles from the mouth of the Savannah River (Georgia) in Wassaw Sound off Tybee Beach from an altitude of approximately 7,200 feet at an aircraft speed of 180–190 knots. The precise weapon impact point is unknown. No detonation occurred, and the B–47 landed safely.

13. MARCH 11, 1958. FLORENCE, SOUTH CAROLINA.

On March 11, 1958 at 3:53 p.m., a B–47E departed from Hunter AFB, Georgia, as the third aircraft in a flight of four enroute to an overseas base. After leveling off at 15,000 feet, the aircraft accidentally jettisoned an unarmed nuclear weapon that impacted in a sparsely populated area six and one-half miles east of Florence, South Carolina. The bomb's HE material exploded on impact, causing property damage and several injuries on the ground. The aircraft returned to base without further incident. No capsule of nuclear materials was aboard the B–47 or installed in the weapon.

14. NOVEMBER 4,1958. DYESS AFB, TEXAS.

A B–47 caught fire on take-off. Three crew members successfully ejected but one was killed when the aircraft crashed from an altitude of 1,500 feet. One nuclear weapon was on board when the aircraft crashed. The resulting detonation of the high explosive made a crater thirty-five feet in diameter and six feet deep. Nuclear materials were recovered near the crash site.

15. NOVEMBER 26, 1958. CHENNAULT AFB, LOUISIANA.

A B–47 caught fire on the ground. The single nuclear weapon on board was destroyed by the fire. Contamination was limited to the immediate vicinity of the aircraft wreckage.

16. JANUARY 18, 1959. PACIFIC BASE.

An F–100 was parked on a reveted hardstand in ground alert configuration. The external load consisted of a weapon (on the left intermediate station) and three fuel tanks (both inboard stations and the right intermediate station). When the starter button was depressed during a practice alert, the external fuel tanks inadvertently jettisoned and caused an explosion and fire. Fire trucks put out the fire in about seven minutes. The capsule was not in the vicinity of the aircraft and was not involved in the accident. There were no contamination or cleanup problems.

17. JULY 6, 1959. BARKSDALE AFB, LOUISIANA.

A C–124 on a nuclear logistics mission crashed on take-off. The aircraft was destroyed by fire, destroying one weapon. No nuclear or high-explosive detonation occurred—safety devices functioned as designed. Limited contamination was present over a very small area immediately below the destroyed weapon. This contamination did not hamper rescue or fire-fighting operations.

18. SEPTEMBER 25, 1959. OFF WHIDBEY ISLAND, WASHINGTON.

A U.S. Navy P–5M aircraft ditched in Puget Sound off Whidbey Island, Washington. It was carrying an unarmed nuclear antisubmarine weapon containing no nuclear material. The weapon was not recovered.

19. OCTOBER 15, 1959. HARDINSBERG, KENTUCKY.

A B–52 and KC–135 collided while refueling at 32,000 feet near Hardinsberg, Kentucky. The instructor pilot and pilot ejected from the B–52, followed by the electronic warfare officer and the radar navigator. The co-pilot, navigator, instructor navigator, and tail gunner in the B–52 and all four crew members in the KC–135 were killed when the two aircraft crashed. The B–52's two unarmed nuclear weapons were recovered intact, although one had been partially burned. No nuclear materials were dispersed.

20. JUNE 7, 1960. McGUIRE AFB, NEW JERSEY.

A BOMARC air defense missile in ready storage condition (permitting launch in two minutes) was destroyed by explosion and fire after a high-pressure helium tank exploded and ruptured the missile's fuel tanks. The warhead was also destroyed by the fire, although the high explosive did not detonate. Nuclear safety devices functioned properly, and contamination was restricted to an area immediately beneath the weapon and, due to run-off from fire-fighting water, to an adjacent area approximately 100 feet long.

21. JANUARY 24, 1961. GOLDSBORO, NORTH CAROLINA.

During a B–52 airborne alert mission, structural failure of the aircraft's right wing led to the separation of two weapons from an altitude of 2,000–10,000 feet. One bomb broke apart upon impact. No explosion occurred. A portion of the other weapon, containing uranium, fell into water-logged farmland but could not be recovered despite efforts to excavate to a depth of fifty feet. The air force subsequently purchased an easement requiring permission for anyone to dig there. There is no detectable radiation and no hazard in the area.

22. MARCH 14, 1961. YUBA CITY, CALIFORNIA.

A B–52, forced to descend to 10,000 feet after its crew pressurization system failed, depleted its fuel prior to a scheduled rendezvous with a tanker. Most of the crew bailed out at 10,000 feet; the commander stayed with the aircraft to 4,000 feet in order to steer it away from populated areas. The two nuclear weapons on board were torn from the aircraft upon impact. The high explosives did not detonate. Safety devices functioned properly and there was no nuclear contamination.

23. NOVEMBER 13, 1963. IGLOO/MEDINA BASE, TEXAS.

An explosion involving 123,000 pounds of high-explosive components of nuclear weapons in an Atomic Energy Commission storage igloo caused minor injuries to three AEC employees. There was little contamination from nuclear components stored elsewhere in the building.

24. JANUARY 13, 1964. CUMBERLAND, MARYLAND.

A B–52D enroute from Westover Air Force Base, Massachusetts, to its home base at Turner Air Force Base, Georgia, crashed approximately seventeen miles southwest of Cumberland, Maryland. The aircraft was carrying two weapons. Both weapons were in a tactical ferry configuration (no mechanical or electrical connections had been made to the aircraft and the switches were in the "SAFE" position). Prior to the crash, the pilot had requested a change of altitude because of severe air turbulence at 29,500 feet. The aircraft was cleared to climb to 33,000 feet. During the climb, the aircraft encountered violent air turbulence that caused structural failure. Both weapons remained in the aircraft until it crashed and were found relatively intact in the approximate center of the wreckage area. Two of five crew members survived after ejecting from the aircraft.

25. DECEMBER 5, 1964. ELLSWORTH AFB, SOUTH DAKOTA.

An LGM 30B Minuteman I missile was on strategic alert an Launch Facility (LF) L–02, Ellsworth AFB, South Dakota. While two airmen were repairing the inner zone security system, one retro rocket in the spacer below the Reentry Vehicle (RV) fired, causing the RV to fall about seventy-five feet to the floor of the silo. Although the RV structure received considerable damage, all safety devices operated properly. The proper warhead arming sequence was not activated; there was no detonation or radioactive contamination.

26. DECEMBER 8, 1964. BUNKER HILL (NOW GRISSOM) AFB, INDIANA.

While taxing during an exercise alert, a B–58 slid off the runway and burned. Portions of the five on-board nuclear weapons burned, and contamination was limited to the immediate area of the crash and subsequently removed.

27. OCTOBER 11, 1965. WRIGHT-PATTERSON AFB, OHIO.

A C–124 was being refueled in preparation for a routine logistics mission when a fire occurred at the aft end of the refueling trailer. The fuselage of the aircraft, containing only components of nuclear weapons and a dummy training unit, was destroyed by the fire. There were no casualties. The resultant radiation hazard was minimal. Minor contamination found on the aircraft, cargo, and cleanup personnel was removed by normal cleaning.

28. DECEMBER 5, 1965. AT SEA, PACIFIC.

An A–4 aircraft loaded with one nuclear weapon rolled off the elevator of a U.S. aircraft carrier and fell into the sea. The pilot, aircraft, and weapon were lost. The incident occurred more than 500 miles from land.

29. JANUARY 17, 1966. PALOMARES, SPAIN.

A B–52 and KC–135 collided near Palomares, Spain, during a routine high-altitude air refueling operation. Four of the eleven crew members survived. The B–52 carried four nuclear weapons. One was recovered intact on the ground, and one was recovered from the sea three months later, after extensive search and recovery efforts. Two of the weapons exploded on impact with the ground, dispersing some radioactive materials. Approximately 1,400 tons of slightly contaminated soil and vegetation were shipped to the United States for storage at an approved site. Representatives of the Spanish government monitored the cleanup operation.

30. JANUARY 21, 1968. THULE, GREENLAND.

A B–52 from Plattsburgh AFB, New York, burned after crashlanding on sea ice seven miles southwest of the runway at Thule AFB, Greenland. Six of the seven crew members survived. The four nuclear weapons carried by the bomber were destroyed by fire. Some radioactive contamination occurred in the vicinity of the crash. Some 237,000 cubic feet of contaminated ice, snow, and water, with crash debris, were shipped to an approved storage site in the United States. Although the extent of contamination was unknown, environmental sampling showed normal readings after the cleanup was completed. Representatives of the Danish government monitored the cleanup operations.

31. SPRING 1968. THE ATLANTIC OCEAN.

Details remain classified.

32. SEPTEMBER 19, 1980. DAMASCUS, ARKANSAS.

During routine maintenance in a Titan II silo, an Air Force repairman dropped a heavy wrench socket off a work platform. The socket struck the missile and caused the pressurized fuel tank to leak. The missile complex and the surrounding area were evacuated, and a team of specialists was called in from Little Rock Air Force Base, the missile's main support base. About eight and one-half hours after the initial puncture, fuel vapors within the silo ignited and exploded. The explosion fatally injured one member of the team. Twenty-one other air force personnel were injured. The missile's re-entry vehicle, which contained a nuclear warhead, was recovered intact. There was no radioactive contamination.